VISIONS OF ENERGY FUTURES

This book examines the visions, fantasies, frames, discourses, imaginaries, and expectations associated with six state-of-the-art energy systems—nuclear power, hydrogen fuel cells, shale gas, clean coal, smart meters, and electric vehicles—playing a key role in current deliberations about low-carbon energy supply and use.

Visions of Energy Futures: Imagining and Innovating Low-Carbon Transitions unveils what the future of energy systems could look like, and how their meanings are produced, often alongside moments of contestation. Theoretically, it analyzes these technological case studies with emerging concepts from various disciplines: utopianism (history of technology), symbolic convergence (communication studies), technological frames (social construction of technology), discursive coalitions (discourse analysis and linguistics), sociotechnical imaginaries (science and technology studies), and the sociology of expectations (innovation studies, future studies). It draws from these cases to create a synthetic set of dichotomies and frameworks for energy futures based on original data collected across two global epistemic communities—nuclear physicists and hydrogen engineers—and experts in Eastern Europe and the Nordic region, stakeholders in South Africa, and newspapers in the United Kingdom. This book is motivated by the premise that tackling climate change via low-carbon energy systems and practices is one of the most significant challenges of the twenty-first century, and that success will require not only new energy technologies, but also new ways of understanding language, visions, and discursive politics. The discursive creation of the energy systems of tomorrow is propagated in polity, hoping to be realized as the material fact of the future, but processed in conflicting ways with underlying tensions as to how contemporary societies ought to be ordered.

This book will be essential reading for students and scholars of energy policy, energy and environment, and technology assessment.

Benjamin K. Sovacool is Professor of Energy Policy at the Science Policy Research Unit (SPRU) at the School of Business, Management, and Economics, University of Sussex, UK. He is also Professor of Business and Social Sciences at Aarhus University in Denmark, where he also directs the Centre on Energy Technologies.

Routledge Studies in Energy Transitions

Considerable interest exists today in energy transitions. Whether one looks at diverse efforts to decarbonize, or strategies to improve the access levels, security and innovation in energy systems, one finds that change in energy systems is a prime priority.

Routledge Studies in Energy Transitions aims to advance the thinking which underlies these efforts. The series connects distinct lines of inquiry from planning and policy, engineering and the natural sciences, history of technology, STS, and management. In doing so, it provides primary references that function like a set of international, technical meetings. Single and co-authored monographs are welcome, as well as edited volumes relating to themes, like resilience and system risk.

Series Editor: Dr. Kathleen Araújo, Boise State University and Energy Policy Institute, Center for Advanced Energy Studies (US)

Series Advisory Board
Morgan Bazilian, Colorado School of Mines (US)
Thomas Birkland, North Carolina State University (US)
Aleh Cherp, Central European University (CEU, Budapest) and Lund University (Sweden)
Mohamed El-Ashry, UN Foundation
Jose Goldemberg, Universidade de Sao Paolo (Brasil) and UN Development Program, World Energy Assessment
Michael Howlett, Simon Fraser University (Canada)
Jon Ingimarsson, Landsvirkjun, National Power Company (Iceland)
Michael Jefferson, ESCP Europe Business School
Jessica Jewell, IIASA (Austria)
Florian Kern, Institut für Ökologische Wirtschaftsforschung (Germany)
Derk Loorbach, DRIFT (Netherlands)
Jochen Markard, ETH (Switzerland)
Nabojsa Nakicenovic, IIASA (Austria)
Martin Pasqualetti, Arizona State University, School of Geographical Sciences and Urban Planning (US)
Mark Radka, UN Environment Programme, Energy, Climate, and Technology
Rob Raven, Utrecht University (Netherlands)
Roberto Schaeffer, Universidade Federal do Rio de Janeiro, Energy Planning Program, COPPE (Brasil)
Miranda Schreurs, Technische Universität München, Bavarian School of Public Policy (Germany)

Vaclav Smil, University of Manitoba and Royal Society of Canada (Canada)
Benjamin Sovacool, Science Policy Research Unit (SPRU), University of
Sussex (UK)

Titles in this series include:

How Power Shapes Energy Transitions in Southeast Asia
A Complex Governance Challenge
Jens Marquardt

Energy and Economic Growth
Why we Need a New Pathway to Prosperity
Timothy J. Foxon

Accelerating Sustainable Energy Transition(s) in Developing Countries
The Challenges of Climate Change and Sustainable Development
Laurence L. Delina

Petroleum Industry Transformations
Lessons from Norway and Beyond
Edited by Taran Thune, Ole Andreas Engen and Olav Wicken

Sustainable Energy Transformations, Power and Politics
Morocco and the Mediterranean
Sharlissa Moore

Energy as a Sociotechnical Problem
An Interdisciplinary Perspective on Control, Change, and Action in Energy Transitions
Edited by Christian Büscher, Jens Schippl and Patrick Sumpf

Transitions in Energy Efficiency and Demand
The Emergence, Diffusion and Impact of Low-Carbon Innovation
Edited by Kirsten E.H. Jenkins and Debbie Hopkins

Visions of Energy Futures
Imagining and Innovating Low-Carbon Transitions
Benjamin K. Sovacool

For more information about this series, please visit: www.routledge.com/Routledge-
Studies-in-Energy-Transitions/book-series/RSENT

VISIONS OF ENERGY FUTURES

Imagining and Innovating Low-Carbon Transitions

Benjamin K. Sovacool

LONDON AND NEW YORK

First published 2019
by Routledge
2 Park Square, Milton Park, Abingdon, Oxon OX14 4RN

and by Routledge
52 Vanderbilt Avenue, New York, NY 10017

Routledge is an imprint of the Taylor & Francis Group, an informa business

British Library Cataloguing-in-Publication Data
A catalogue record for this book is available from the British Library

Library of Congress Cataloging-in-Publication Data
Names: Sovacool, Benjamin K., author.
Title: Visions of energy futures : imagining and innovating low-carbon
transitions / Benjamin K. Sovacool.
Description: Abingdon, Oxon ; New York, NY : Routledge, [2019] |
Series: Routledge studies in energy transitions | Includes bibliographical
references and index.
Identifiers: LCCN 2018048639 (print) | LCCN 2018059719 (ebook) |
ISBN 9780367135171 (Master) | ISBN 9780367111991 (hbk) | ISBN
9780367112004 (pbk) | ISBN 9780367135171 (ebk)
Subjects: LCSH: Clean energy industries–Case studies. | Energy
industries–Technological innovations. | Energy policy. | Carbon dioxide
mitigation. | Power resources–Political aspects.
Classification: LCC HD9502.5.C542 (ebook) | LCC HD9502.5.C542 S68
2019 (print) | DDC 333.79/4–dc23
LC record available at https://lccn.loc.gov/2018048639

ISBN: 978-0-367-11199-1 (hbk)
ISBN: 978-0-367-11200-4 (pbk)
ISBN: 978-0-367-13517-1 (ebk)

Typeset in Bembo
by Swales & Willis, Exeter, Devon, UK

Printed and bound in Great Britain by
TJ International Ltd, Padstow, Cornwall

CONTENTS

FIGURES

TABLES

ABOUT THE AUTHOR

Benjamin K. Sovacool, Ph.D, is Professor of Energy Policy at the Science Policy Research Unit (SPRU) at the School of Business, Management, and Economics, part of the University of Sussex in the United Kingdom. There he serves as Director of the Sussex Energy Group and Director of the Center on Innovation and Energy Demand. He is also Director of the Center for Energy Technologies and Professor of Business and Social Sciences in the Department of Business Development and Technology at Aarhus University in Denmark. Professor Sovacool works as a researcher and consultant on issues pertaining to energy policy, energy security, climate change mitigation, and climate change adaptation. More specifically, his research focuses on renewable energy and energy efficiency, the politics of large-scale energy infrastructure, designing public policy to improve energy security and access to electricity, and building adaptive capacity to the consequences of climate change. Professor Sovacool is the author of numerous refereed articles, book chapters, and reports, including solely authored pieces in *Nature* and *Science*. He is the author, co-author, editor, or co-editor of 22 books, including *Climate Change and Global Energy Security*, *Energy Poverty*, *Global Energy Justice*, *The Political Economy of Climate Change Adaptation*, *Fact and Fiction in Global Energy Policy*, and *Enabling the Great Energy Transition*. Former U.S. President Bill Clinton, the Prime Minister of Norway—Gro Harlem Brundtland, and the late Nobel Laureate—Elinor Ostrom have endorsed his books. Additionally, Professor Sovacool is the founding Editor-in-Chief for the international peer-reviewed journal *Energy Research & Social Science*, and he sits on the Editorial Advisory Panel of *Nature Energy*.

ACKNOWLEDGMENTS

This book is special: the idea for it began more than a decade ago and it took the author more than eight years, multiple methods, numerous projects, and collaborations with excellent colleagues to make it happen. As such, the author is grateful to the U.S. National Science Foundation for grants SES-0522653, ECS-0323344, and SES-0522653, the Research Councils United Kingdom (RCUK) for Energy Program Grant EP/K011790/1, and the Danish Council for Independent Research (DFF) for Sapere Aude Grant 4182-00033B, which have supported elements of the work reported here. Any opinions, findings, and conclusions or recommendations expressed in this material are those of the author and do not necessarily reflect the views of the U.S. National Science Foundation, RCUK Energy Program or the DFF. This project has also received funding from the European Union's Horizon 2020 research and innovation programme under grant agreement No 730403 "Innovation pathways, strategies and policies for the Low-Carbon Transition in Europe (INNOPATHS)". The content of this deliverable does not reflect the official opinion of the European Union. Responsibility for the information and views expressed herein lies entirely with the author(s).

Lastly, some of the chapters in this book build on and extend arguments originally presented in the following articles, in alphabetical order:

- Goldthau, A., and B. K. Sovacool. 2016, November. Energy Technology, Politics, and Interpretative Frames: Shale Gas Fracking in Eastern Europe. Global Environmental Politics 16 (4): 50–69.

- Hielscher, S., and B. K. Sovacool. 2018, September. Contested Smart and Low-Carbon Energy Futures: Media Discourses of Smart Meters in the United Kingdom. Journal of Cleaner Production 195: 978–990.
- Rafey, W., and B. K. Sovacool. 2011, August. Competing Discourses of Energy Development: The Implications of the Medupi Coal-Fired Power Plant in South Africa. Global Environmental Change 21 (3): 1141–1151.
- Sovacool, B. K., and B. Brossmann. 2010, April. Symbolic Convergence and the Hydrogen Economy. Energy Policy 38 (4): 1999–2012.
- Sovacool, B. K., and B. Brossmann. 2013, June. Fantastic Futures and Three American Energy Transitions. Science as Culture 22 (2): 204–212.
- Sovacool, B. K., and B. Brossmann. 2014, September. The Rhetorical Fantasy of Energy Transitions: Implications for Energy Policy and Analysis. Technology Analysis & Strategic Management 26 (7): 837–854.
- Sovacool, B. K., and D. J. Hess. 2017, October. Ordering Theories: Typologies and Conceptual Frameworks for Sociotechnical Change. Social Studies of Science 47 (5): 703–750.
- Sovacool, B. K., J. Kester, L. Noel, and G. Zarazua de Rubens. Contested Visions and Sociotechnical Expectations of Electric Mobility and Vehicle-to-Grid Innovation in Five Nordic Countries. Environmental Innovation and Societal Transitions, in press 2019.
- Sovacool, B. K., and M. V. Ramana. 2015, January. Back to the Future: Small Modular Reactors, Nuclear Fantasies, and Symbolic Convergence. Science, Technology, & Human Values 40 (1): 96–125.
- Sovacool, B. K., and S. V. Valentine. 2012. The National Politics of Nuclear Power: Economics, Security, and Governance (London: Routledge).

Many thanks to the editors and peer reviewers at these journals, and especially my co-authors Brent, William, Scott, Ramana, Andreas, David, Sabine, Johannes, Lance, and Gerardo for helping refine and improve my thinking on the topic.

1

INTRODUCTION

Visions and futures in the study of low-carbon energy systems

> The human condition can almost be summed up in the observation that, whereas all experiences are of the past, all decisions are about the future. The image of the future, therefore, is the key to all choice-oriented behavior. The character and quality of the images of the future which prevail in a society are therefore the most important clue to its overall dynamics.
>
> Kenneth E. Boulding, 1973 In F. L. Polak. The image of the future. Amsterdam/London/New York: Elsevier Scientific. 1973. p. v.

> All the great empires of the future will be empires of the mind.
>
> Sir Winston Churchill, 1953 In Edward Hedican. Anthropology in the future. Social Anthropology: Canadian Perspectives on Culture and Society. Canadian Scholars Press. 2012. p. 237.

As messengers Boulding and Churchill opine, how we think about the future can both reveal fundamental aspects of the human condition, and also motivate exploration and intellectual empire building. Visions of the future have a long history inspiring humanity to create a better tomorrow. Fred Polak, a sociologist who practically founded the field of future studies, goes as far as to argue that the heights of classical civilization, Judaic culture, Islamic culture, the Renaissance and the Enlightenment, and the early industrial era were all preceded by daring imaginative leaps forward and new visions of human possibility.[1] Although one can debate his logic, for Polak, culture itself can be defined by its vision of the future. "Thinking about the future," he concluded, "is not only the mightiest lever of progress but also the condition of survival."[2] Something known as "Quigley's Law" supposes that "successful societies are defined by their readiness to allow consideration of the future to determine today's choices."[3] As the epitaph above suggests, Elise Boulding and Kenneth E. Boulding add that at an even more basic level, unless we think we know something about the future, decisions are impossible; all decisions involve choices among images of alternative futures.[4] To Kenneth E Boulding in particular, "all decisions are made about imaginary futures, not about real ones"[5]

Visions of the future can also play a key role in the innovation and research process itself. Fujimura's work has demonstrated how genomic scientists use "future imaginaries" to mobilize financial support for their research,[6] and Van Lente has similarly shown that fantastic expectations of technology can motivate engineers and designers to initiate projects.[7] Jasanoff and Kim write about "socio-technical imaginaries" operating behind nuclear research in South Korea and the United States, and point out that national "imaginations can penetrate the very designs and practices of scientific research and technological development."[8] Mads Borup and colleagues argue that expectations are of great importance for the development of technologies as they stimulate, steer and coordinate actors as diverse as designers, managers, investors, sponsors, and politicians. As they go on to write:

> Future-oriented abstractions are among the most important objects of enquiry for scholars and analysts of innovation. Such expectations can be seen to be fundamentally "generative," they guide activities provide structure and legitimation, attract interest and foster investment. They give definition to roles, clarify duties, offer some shared shape of what to expect and how to prepare for opportunities and risks. Visions drive technical and scientific activity, warranting the production of measurements, calculations, material tests, pilot projects and models. As such, very little in innovation can work in isolation from a highly dynamic and variegated body of future-oriented understandings about the future.[9]

Offering more support, Cynthia Selin adds that "the expectations, hopes, fears, and promises of new technologies are not set apart from, nor layered on top of scientific and technological practices but are, rather, formative elements of innovation,"[10] and Nightingale calls technological optimism and fantasy an elemental part of the "cognitive" dimension of innovation.[11] Technological visions and fantasies can even become exclusionary and self-replicating, convincing those that do not share them to leave a project or disciplinary field entirely. To all of these disparate types of thinkers, "the future is what matters in the present."[12]

For perhaps these reasons, visions of the energy future in particular have become a powerful force in the construction (and deconstruction) of energy and climate scenarios, forecasts, and analysis. Outside of the research community, business analysts, regulators, titans of industry and inventors (among others) continually devote a significant amount of effort towards developing, deploying, and even negating futuristic narratives and images for political and economic ends. In the public domain, users, consumers, citizens, and the media also frequently invent, modify, circulate, and/or resist such narratives.[13]

Therefore, fantastic, visionary, and delightful narratives can capture public imaginations and solicit political support for low-carbon technologies. They can facilitate Churchill's "empire building of the mind." One can even see appeals to imagination and fantasy when they visit local museums or funding agencies such as the European Commission, as Figure 1.1 indicates. Visions and fantasies therefore have relevance for all those concerned about energy technology choice, innovation, commercialization, and energy and climate policy decisions.[14,15,16]

Notwithstanding their prominence, little academic work has attempted to engage the topic of visions and fantasies empirically or theoretically in a systematic or comparative manner, or connected them to pressing policy concerns such as low-carbon transitions. To address this gap, this book seeks to examine the visions (and fantasies, frames, discourses, imaginaries, and expectations) associated with six state-of-the-art energy systems—nuclear power, hydrogen fuel cells, shale gas, clean coal, smart meters, and electric vehicles—playing a key role in current deliberations about low-carbon energy supply and use. The book's methodology is based on extensive original data or analysis including semi-structured interviews, media content analysis, and systematic reviews.

Why these six innovations? To offer a balance between supply side options (gas, coal, nuclear) and decentralized or even end-use options (fuel cells, smart meters, electric vehicles), as well as a mix of fuels (some renewable, some fossil fueled, and fission) and scales (household, industrial, commercial). Moreover, although we will see how some have deeper historical origins than commonly articulated, all of them are relatively new, and at early stages of commercialization or diffusion, meaning that critically examining them can influence actual deployment and innovation trajectories and ultimately consumer acceptance (or rejection). There was lastly an effort at novelty, to break new ground and not study energy technologies that have received the bulk of visions or discourse scholarship so far, notably biofuel and biomass,[17,18,19,20,21,22] hydropower,[23,24,25,26] wind energy,[27,28,29,30,31] and solar energy[32,33,34,35,36] (which is why these types of energy systems are not examined here).

Ultimately, the book is motivated by the premise that tackling climate change via low-carbon energy systems (and practices) is one of the most significant challenges of the twenty-first century, and that success will require not only new energy technologies but also new ways of understanding language, visions, and discursive politics. In doing so, the book unveils what the future of energy systems could look like and how their meanings are produced, often alongside moments of contestation. The discursive creation of the energy systems of tomorrow are often propagated in polity, hoping to be realized as the material fact of the future but processed in conflicting ways with underlying tensions as to how contemporary societies ought to be ordered (and disordered). Visions thus nestle in a critical space between the possible and impossible, the emergent and divergent, and the material and corporeal.

Definitions and terms

Admittedly, some of the core terms and concepts in the book—vision, discourse, narrative, expectation, storyline, fantasy—interrelate and overlap. Although I offer

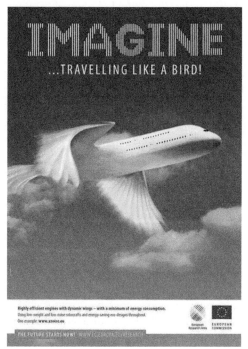

FIGURE 1.1 Imaginary visions about energy and transport in London and Brussels, 2017. (a) Top panel: At the London Transport Museum, Covent Garden, London, October 2017. (b) Bottom panel: At the Research and Innovation Offices of the European Commission, Brussels, Belgium, May 2017

Source: Author

varying definitions throughout the chapters, the term "vision" is broadly meant to be an umbrella concept covering "a description of what could occur in the near-term, mid-term, or long-term future." At the core of analyzing a vision is an assessment of the symbolic aspects of communication, highlighting the relationship between what passes for scientific reality or engineering capability, and things that mediate that reality such as stories, expectations, and visions.[37,38,39,40,41] In other terms, articulated visions are simultaneously rational and allegorical, reflecting both a penchant for storytelling and constructing myths as much as logical reasoning or rational action.

That said, my use of the term "fantasy" is more precise: it refers to "a storyline that captures the human need to experience and interpret drama." For Bormann (see Chapter Three on "symbolic convergence" for more), "fantasy" refers to the way that communities of people share their social reality, a creative interpretation of events that fulfill a psychological and rhetorical need.[42] It is *not* to be mistaken for something that is imaginary (like ghosts or aliens), pejorative (signifying someone that is insane), sexual (as in erotic fantasies), or distinct from reality (like *Lord of the Rings*). Fantasies imply that the boundary between technical, rhetorical, and policy issues is a porous one, and show how actors become enrolled in particular storylines and plots.[43,44]

We are often accustomed to viewing rhetoric or discourse as an ornament of speech—sometimes even pretentious, superficial, or unnecessary. This view, however, ignores that many of the Greeks and their heirs saw rhetoric not merely as dramatic performance but as an indispensable means of communicating truth and shaping reality. Furthermore, studying visions reveals fundamental patterns of human reasoning, and how humans communicate their thinking to others. Berkout compellingly notes that visions can play at least five different, important and active roles.[45] They can:

- Map a possibility space by identifying a realm of plausible alternates and the means for reaching them;
- Offer a heuristic device for revealing the specific problems that need to be resolved in order for a vision to be realized;
- Enable the identification of stable frames for target setting and monitoring progress;
- Specify metaphors and relevant symbols, narratives, or moralities that bind together different stakeholder groups;
- Bring together capital, knowledge, networks, skills, and other resources so that action can be coordinated and focused.

Because the technological landscape is always changing, visions such as these can shape how we think about energy and therefore shape the energy future. Furthermore, and critically, deconstructing visions and discursive tactics can both offer a diagnostic tool (learning from previous visions) and also reveal the vested and even hegemonic interests and power relations underlying them.[46]

Here, Churchill's quote takes on another meaning: if visions are about intellectual empire building, then they can also be about intellectually colonizing the future for one's own ends.

Topical, empirical, and conceptual contributions

In proceeding to analyze energy visions and fantasies, the book refuses to approach the issue within disciplinary boundaries. It differs from previous scholarship—and is novel—in four ways.

First, the book investigates *low-carbon transitions*. The book is inherently cross-technological, looking at low-carbon technologies at a variety of scales and purposes across the domains of electricity supply, transport and mobility, industry, and household energy use. As Cozen et al. write:

> Energy transitions—including international and national policy shifts, rapid technological changes, and public decision-making—address some of the most important sustainability challenges facing society. A sustainable future depends on how we think about, communicate about, and use energy.[47]

Since our modern civilization is built on the ubiquity of energy fuels and services, low-carbon energy transitions provide insight into not only the history (and possible future) of energy flows but also questions of equity (how energy resources are distributed), futurity (what energy systems we may be transitioning to over time), and values (the things we deem important enough to put energy services to use achieving). Put another way, the discursive tactics over energy are symbolic of future global struggles to simultaneously expand access to energy services and minimize environmental degradation. As Sovacool and Brossmann note, "how people imagine energy technologies and their futures is clearly important to understanding how and why people invest in them, financially, personally, professionally, and otherwise, and it is thus a critical social facet of energy transitions."[48]

Second, the book is *spatially and temporally comparative*, examining several specific geographic contexts where such visions and narratives play out: globally in two epistemic communities (nuclear physicists and hydrogen engineers), nationally in South Africa and the United Kingdom, and regionally in Eastern Europe and the Nordic countries. I also present an examination of four historical transitions (steam engines, cars, dams, and nuclear reactors) in this introductory chapter covering a time period of more than 150 years.

Third, the book is *data-driven and multi-method*. It draws on original and unique datasets across each chapter. Chapter Two on "Technological utopianism: Small modular reactors and the physics community" relies on content analysis of nuclear physics journals as well as International Atomic Energy Agency technical reports. Chapter Three on "Symbolic convergence: Hydrogen fuel cells and the engineering community" relies on both research interviews as well as a content analysis of public media and the academic literature. Chapter Four on "Techno-logical frames: The interpretive flexibility of shale gas in Eastern Europe" relies on semi-structured research interviews across Poland, Bulgaria, and Romania. Chapter Five on "Discursive coalitions: Contesting clean coal in South Africa" utilizes a content analysis of project documents and public media surrounding the Medupi Coal-fired Power Plant. Chapter Six on "Sociotechnical imaginaries: Smart meters and the public in the United Kingdom" draws on a systematic content analysis of 11 years of newspaper media coverage. Chapter Seven on "Expectations: Electric mobility and experts in the Nordic region" draws on semi-structured expert interviews across Denmark, Finland, Iceland, Norway, and Sweden.

Fourth and finally, the book is *trans-theoretical*, engaging and applying six different conceptual approaches or heuristics often used in isolation. Concep-tually, the book analyzes its technological case studies with emerging concepts from various disciplines: utopianism (history of technology), symbolic conver-gence (communication studies), technological frames (social construction of technology), discursive coalitions (discourse analysis and linguistics), imaginaries (science and technology studies), and expectations (innovation studies, future studies). It then draws from these cases and frameworks to discuss dimensions and dichotomies of energy futures in the concluding chapter.

In proceeding as such, the book takes a critical step towards a much-needed rhetoric of low-carbon energy technology. Most assessments of communication and science and technology have tended to focus on one of two themes. On the one theme, there are those authors that focus on the rhetoric of science. For example, the now classic texts of Robert K. Merton,[49] Thomas Kuhn,[50,51] and Derek De Sola Price[52] have shown how scientists set rhetorical boundaries that demarcate their disciplines from other professions. Later works have focused intensely on how scientists and engineers have used such boundary setting to enhance their credibility and attract funding for research projects.[53,54,55,56] Theo-dore Porter[57] and Ulrich Beck[58] have done an excellent job analyzing the rhetorical tropes and lines of argument, such as quantification and risk assessment, commonly used in scientific discourse, and Alan Gross has documented many of the logical and stylistic choices utilized by different scientific communities.[59] Others have focused on the common lines of argument made by key stakeholders in debates on renewable energy technologies,[60] and Bruno Latour[61] and Harry Collins[62] have shown how scientific research depends intricately on the social networking and persuasive skills of its practitioners.

On the other theme, there is an ever-growing body of scholarship focusing on how technologies change the content and form of communication. To cite just a few prominent and popular examples, Elizabeth Eisenstein argued that the shift from script to print significantly influenced the Renaissance, the Reformation, and the rise of modern science.[63] Marshall McLuhan believed that devices such as radio and television changed the nature of communication by creating an interconnected "global village."[64] Neil Postman suggested that technologies such as the telegraph, photography, and television greatly altered the form of communication by de-contextualizing it from local events and assailing audiences with hoards of irrelevant information.[65] Jeremy Rifkin posited that the internet will dramatically alter fundamental notions of property ownership and capital accumulation.[66]

The above two themes demonstrate that communication practices play an important role in scientific research, and that technologies can alter both the substance and style of communication. However, the first school glosses over the fact that technology can exert as much influence on social structures as science, and second that changes in the content or substance of communication can also exert considerable influence on technological choices. As Bijker put it when talking about the interaction between technology and society:

> Purely social relations are to be found only in the imaginations of sociologists, among baboons, or possibly on nudist beaches; and purely technical relations are to be found only in the sophisticated reaches of science fiction. The technical is socially constructed, and the social is technically constructed—all stable ensembles are bound together as much by the technical as by the social. Where there was purity, there is now heterogeneity. Social classes, occupational groups, firms, professions, machines—all are held in place by intimately linked social and technical means ... . Society is not determined by technology, nor is technology determined by society. Both emerge as two sides of the sociotechnical coin.[67]

Given that technical and scientific issues can never be divorced from social ones, and that technologies themselves interact in a mutually constitutive way with society and communicative practices, technological systems can greatly influence the rhetorical choices available to a given community or culture. Exploring this relationship in terms of low-carbon technologies offers a more nuanced look at how visions of alternative energy futures are constructed and maintained. It also offers an intricate look at the power politics involved in the processes of vision-making and in materializing these futures.

The coevolution of energy visions and technologies: four histories

In exploring the connection between visions and low-carbon energy transitions, scholarship emerging from science and technology studies, psychology, and communication studies illustrates how fantasies and values coevolve with technology and social expectations over time.

For example, studies by Olson,[68] Segal,[69,70] and Berkout[71] have explored the relationship between technology and fantasy, and they tend to agree that effective depictions of future technological visions usually share four common themes. They are concrete enough to be applicable to the real world. They remain critical of contemporary time. They provide compelling arguments for social change.

They argue that the impact of technological utopia would be so great as to render previous transitions irrelevant (although most never come close to fulfilling that vision).

Geels and Smit examined expectations and projections about future technologies from the 1880s to the 1990s and asked why things such as teleconferencing and flying cars, advocated as complete replacements for face-to-face interaction and transport respectively, never did so.[72] They found that expectations about technology were often "off" because they were made by biased vendors and promoters. Many individuals forecasting trends viewed the future as an extrapolation of the present, and speculated that new technologies would completely replace older ones. In fact, the future is notoriously unpredictable and transitions are partial (substitution from one system to the next rarely happens completely), evident in our brief historical case studies below on hydroelectricity and nuclear power.

Bazerman shows that new technologies (such as electric lighting) have long been understood in terms of narratives and presented through a variety of rhetorical strategies.[73] He investigated the rhetoric and visions used to create meaning and value for the emergent technology of electric lighting, tracing Thomas Edison's considerable ability to shift registers when dealing with different groups. Edison was not only a gifted scientist and a talented engineer; he was an accomplished social actor and communicator, who knew how to sell his new technology to newspaper reporters, how to encourage his inventive team to resolve problems, and how to convince investors to make and sustain their financial commitment to his research. Edison's strategy for promoting electric lights demonstrates a co-evolution of technical performance with systems of meaning and strategies of communication.

These depictions may be rooted in the nature of language itself. Efforts to prepare society for technological visions are attempts at persuasion reliant on what Kenneth Burke[74] calls "the use of language as a symbolic means of inducing cooperation in beings that by nature respond to symbols." This linguistic choice establishes both terms and boundaries, so that terminology becomes not only "a *reflection* of reality, by its very nature as a terminology it must be a *selection* of reality; and to this extent it must function also as a *deflection* of reality."[75] Those who advance a rhetorical vision naturally shape and limit the scope of how the vision is discussed.

Put another way, all individuals suffer from varying degrees of "trained incapacity" and "occupational psychosis," related terms that describe how training and professional occupations prepare people to see the world in certain ways while simultaneously blinding themselves to other perspectives. Some have even termed this a "reality distortion field,"[76] when people suspend disbelief in favor of a stronger, more positive vision about the future, a particular technology, or a charismatic leader. Human motivation can push advocates to seek perfect, homogenizing and even hegemonic visions, and since advocates frame problems in ways that advance their solutions as perfect choices, audiences remain limited in the interpretations available to them. Thus new technologies become evaluated primarily as solutions to existing problems (as defined by the technology's advocates) without much consideration of alternatives.[77] Additionally, because terminology selects, reflects and deflects, much of the "reality" associated with a particular vision actually constitutes a logical implication of the linguistic frames selected by advocates, not of the technology itself.

As these rhetorical frames enter the mainstream, they can take on lives of their own. Bormann's[78] symbolic convergence theory (examined in a later chapter) shows how these visions play out in fantasy themes, or smaller units of analysis that help people tap into larger narratives jointly produced by a community. Like "gravity," fantasy has evolved into a force humans need to explain and interpret their experiences. Frequently employing dramatis personae, these narratives develop plot lines in which characters in everyday situations reduce complexities into meaningful stories people can share, create and elaborate.

In their own meta-analysis of the literature on sociotechnical expectations, Brown and Michael argue that most visions fall into two distinct "interpretive registers": retrospecting prospects, and prospecting retrospects. Retrospecting prospects refers to the way the future was once represented long ago, and it entails a process of recollecting past futures or peoples' memories of the future. Prospecting retrospects refers to what people do in the present, the uses people have for memories by redeploying them to engage with the future.[79] Essentially, retrospecting prospects look backwards, and prospecting retrospects look forwards.

As all of the chapters to come focus on different degrees of prospecting retrospects, here I thought it insightful to briefly capture four retrospecting prospects. To both underscore the value of themes such as these in examining visions, and to offer context to the chapters to come, this section examines (drawn from previous research[80,81]) four previous visions of an energy future: steam engines, cars, dams, and nuclear reactors.

Steam engines: 1850s to 1970s

The human drive to have some control over our lives or the vagaries of nature is widespread. Even in the most natural setting imaginable, early humans used fire for both heat and cooking, and in so doing separated themselves from a completely natural environment. While it would be inaccurate to argue that energy transitions *created* desires to master nature or provide structure to civilizations, it is clear that the arguments in favor of particular energy revolutions are built around the twin concepts of control and structure, and that those concepts become modified to fit the rhetorical and actual potentials of a new energy source. The coal–fueled steam engine arose at a time of rapid growth in industrial systems, and when science and technology were intimately connected to notions of progress.

One huge advantage of the steam engine was that it offered *control*. It was an ingenious device that produced large amounts of power by converting heat energy from a burning fuel into mechanical energy. It offered a source of controlled force and motion more powerful than human and animal muscles, yet also more predictable than wind or water. The steam engine "tipped the balance" in giving us potent additional ways to perform work; for the first time energy could be concentrated in a fuel, usable energy stored in a form that could be stockpiled, transported, and released in consolidated form at human will.[82]

Control also provided the ability to create new *social and technological orders*. The steam engine enabled economic or productive activity that, unlike humans and animals, never loitered or lagged, and did their task until it was done.[83] The steam engine not only created systems of production that could replace humans and animals but allowed for the establishment of new perspectives on the expected contributions of living workers. By the late 1900s, the "vehement competition" between humans and steam engines was principally over, because "the steam engine is a competitor which drives him easily out of the market."[84] This provided an impetus for yet another shift, this time toward a service economy.

The steam engine was also regarded as the herald of *progress* and as such precipitated *national pride*. In Great Britain, where the steam engine first appeared, it was admired as a harbinger of enlightenment. One British writer even claimed that just as man was the noblest work of God, so the steam engine was the noblest work of man.[85] Across the ocean in the United States, people celebrated the centenary of their independence in Philadelphia in 1876, and the central symbol of American progress was a large, 1,400 horsepower Corliss steam engine standing 40 feet above a platform, and supplying power to the machinery on display. To the spectators at hand, the steam engine symbolized "American civilization" and the "enterprising American spirit," with one observer remarking that "one cannot fail to utter his pride and content ... it is still in these things of iron and steel that the national genius most freely speaks."[86] The implications were similar in China. Steam engines were introduced there during the First Opium War (1839 to1842), where British superiority in guns and steam ships were central to the defeat of the Qing Dynasty. Although the Chinese developed a desire for both, it took a generation for them to develop steam engines, primarily due to misinterpretations of fundamental principles (attributing the power to fire, not steam), a lack of machine tools, and a reliance on an oral tradition that was not disposed in favor of precise technological drawings.[87] Regardless of the delay, the impact of steam engines was no less momentous, having been credited with the creation of a new Chinese work ethic, a technological revolution, the shift from wood to machines, and a strict division of labor.

Though the social costs of steam engines in terms of pollution, boiler explosions, and mining disasters associated with coal loomed large at the time of their emergence, these things seemed acceptable in exchange for rapid economic development. The steam engine became synonymous with progress and efficiency. Railroads and marine vessels were made faster, homes and public buildings could afford inexpensive heating and lighting. Smokestacks bellowing pollution into the sky even became prized as industrial emblems.[88] Companies and manufacturers flaunted sooty plumes on advertisements, letterheads, and stock certificates, or as one author put it, "effluence signified affluence."[89] Steam power, steamships, locomotives, and miniature steam generators were seen as "sublime" in their ability to solve social ills, lighten the toil of workers and housewives, provide faster and cleaner forms of transport, and revolutionize food production on the farm.[90]

The appeal of the steam engine was only enhanced when it was configured as a source of electricity from a power plant or fleet of power plants. Journalist Chester T. Cromwell wrote in a 1926 edition of the *Saturday Evening Post* that when used to generate electricity, the steam engine provided "nine slaves for each citizen."[91] He argued that steam-produced electricity brought with it "a revolution compared with which Marxian Socialism or Russian Bolshevism would be puny and futile, a mere dance in the wind," making life in America a "great romantic adventure of beauty and comfort for the average person."[92] Crowell concluded that the steam engine enabled "mechanical slaves" to "replace human drudgery," that they liberated "elbows and the enfranchisement of brains," and that in the era of the steam-based power plant, "knowledge is horse power and kilowatt hours."[93]

Reporters were not the only ones to exalt the awesome power of the steam-turbine power plant. A single steam turbine generating unit built in 1925 was praised for producing the energy equivalent work of 900,000 people. Owen D. Young from the manufacturer General Electric interpreted its larger meaning as follows:

> This is the way America must solve her problem of maintaining higher wages than any other country in the world and at the same time keeping her goods competitive in foreign markets. We must put more energy [sic] back of the worker, in order that he may be a director of power rather than a generator of it … I want to see this art not only run the giant industries of the cities, but I want it also to be so humble and true in its social service that we shall banish from the farmers' homes the drudgery which in the earlier days killed their wives. We have come here to dedicate a power plant-an instrument of utility. Is it only that? Perhaps it is a temple.[94]

Young's ebullience for steam power is also reflected in the fact that in the decades following the 1920s—especially after the end of World War II up until the 1960s and 1970s—steam turbines generated 83 percent of electric power in the United States and roughly 75 percent of electric power worldwide.[95]

Gasoline powered automobiles: 1910 to present

Control, social order and progress were instrumental visions in the advent of the steam engine economy. Although the transition to gasoline powered automobiles was perhaps more personal than the advent of the steam engine, similar themes prevailed. Gasoline automobiles provided individuals with more control over their daily lives, allowed them to redefine their place in the social order, and in doing these things, created a sense of progress. This is most evident in the society that was most reliant on gas automobiles—the United States.

Many citizens initially perceived gasoline-powered automobiles as emblematic of mobility, individualism, and progress.[96] Throughout the first decades of the twentieth century, an abundance of literature described gasoline powered car travel as a revival of pre-modern experiences. The greater range of automobiles and their independence from urban areas helped fulfill the desire for a lost frontier. After 1910, gasoline automobiles began to enjoy mass appeal as an instrument of nostalgic recreation. In contrast to the standardization and centralization of rail travel, travelers viewed gas powered automobiles as liberating because drivers became their own "engineers." Free of tracks, motorists determined their own routes and reclaimed the pioneer's view. Free of trains, drivers did not compete for comfortable seats or control over window shades. Free of strangers, motorists no longer worried about haughty conductors, tip-hungry porters, or potential thieves.[97] These messages, designed to convince people to buy cars, were clearly oriented at a sense of *control*.

Gasoline powered vehicles redefined the existing *social order* by promising a friendlier vision of industrialization. While train travelers saw firsthand the economic and environmental dislocation created by modernization and industrialization (such as factories, pollution, warehouses, and slums), car travelers could enter cities by the best streets or avoid cities all together. Gasoline vehicles also promoted a sense of camaraderie, democracy, and meritocracy. In the United States, broken down motorists helped each other out, exchanged information, and experienced America together.[98] Such experiences included camping, an image popularized by Henry Ford and his infamous nature outings with Thomas Edison and Harvey Firestone. Finally, drivers praised gasoline motoring for restoring old fashioned family solidarity. Families could spend six or eight hours together in the cramped, but intimate, space offered by the gasoline automobile. Gasoline engine automobiles and trucks succeeded because they offered a vision of America that resonated strongly with the values and beliefs of American consumers.[99]

The gasoline automobile was so influential that it evolved symbiotically with urban planning and other forms of infrastructure. During the 1920s and 1930s engineers perfected (and planners adopted) various devices that were part of a high speed system of motorized transportation—grade separation of highway from city streets, traffic circles, divided dual highways, and synchronized stop lights. New designs in bridges as well as the first underwater tunnel which went under the Hudson River were invented. The economic opportunities created by automobiles stimulated a land boom and radically inflated property values near roads, thus creating more incentives for roads. These changes to urban form created almost deterministic necessities of how cities should function, all rooted on auto-centric transportation.[100] Even the mass unemployment, malnutrition and disease of the Great Depression did not slow this trend, as it furthered commitment to the automobile though massive public works projects and road building.[101]

As cars continued to develop, they created a sense of *national pride* for nations around the world. The reasons are somewhat obvious. On an individual level, cars not only make transportation easier but have become associated with economic accomplishment. On a national level, the industry needed to produce them generates significant jobs for automobile production, and also for parts and service. About 14 million jobs in the United States are tied to the modern automobile industry, if inclusive of manufacturing and servicing, supplying fuel, and building or maintaining roads. This makes automotive transportation a key part of the economy, with 20 percent of national retail sales being automobiles, and vehicles representing 20 percent of the entire manufacturing sector's output.[102]

Additionally, motorized vehicles make economic transactions of most types far easier. These arguments are central to the vision of automobile manufacturers. The International Organization of Motor Vehicle Manufacturers argues that the production of automobiles is a key sector of the economies of practically every major country on the planet. As for their vision of the global impact of automobiles, their arguments are hauntingly familiar. They observe:

> Automobiles are a liberating technology for people around the world. The personal automobile allows people to live, work and play in ways that were unimaginable a century ago. Automobiles provide access to markets, to doctors, to jobs. Nearly every car trip ends with either an economic transaction or some other benefit to our quality of life.[103]

The size of the global industry is therefore formidable. In 2017, 47 manufacturers, ranging in size from Rongcheng Hutai Motor to Toyota, built more than 73 million cars and commercial vehicles.[104] Those vehicles are built in more than 40 nations, from Argentina to Uzbekistan.

And the prospects of continued reliance on cars are extremely likely. In 1900, there was roughly one automobile per 10,000 Americans, a number that climbed to 8,417 per 10,000 by 2008.[105] That market seems saturated, but Asia is experiencing explosive growth. Between 1998 and 2008, China's per capita car figures rose from 90 to 357 per 10,000 people, India almost doubled from 76 to 132, and Indonesia almost tripled from 125 to 347. While those percentages of people owning cars are miniscule by American standards, their rapid increase suggest huge growth potential. Couple that with the fact that these are three of the four largest populations on the planet, and the global implications become staggering.[106] Transportation planners expect the world to have two *billion* gasoline powered automobiles by 2030, twice the number than currently exists.[107]

Hydroelectric dams: 1890 to 2010

Hydroelectricity fused visions of cheap energy powering the prospect of commercial and urban development with mastery over nature, utopian social order, and national pride. Politicians and developers believed that hydroelectric dams would make continuous, free, and infinite supplies of power, essentially making electricity possible for the masses.[108]

At the turn of the twentieth century, engineers and utility executives praised hydroelectricity for *control over nature*, for its ability to produce energy without combusting fuel and for "putting nature to work for mankind," earning it the nicknames "white coal," "clean coal," and "super power."[109,110] Geologist and explorer Jon Wesley Powell wrote in 1890 that "conquered rivers are better servants than wild clouds."[111] One article for the *Architectural Forum* written in the late 1890s declared that "there is a feeling of grandeur and of poetry and of beauty in the orderly assembly of this modern, efficient and economical equipment," describing the powerhouse of a hydroelectric dam.[112] H.G. Wells, author of *War of the Worlds*, was so inspired by the hydroelectric station at Niagara Falls that he wrote:

> The dynamos and galleries of the Niagara Falls Power Company impressed me far more profoundly than the Cave of the Winds; are, indeed, to my mind, greater and more beautiful than the accidental eddying of air beside a downpour.[113]

It was during this decade of 1890 to 1900 that hydroelectric dams became, in the words of historian David E. Nye "symbols of both technological progress and economic prosperity."[114] Hydroelectric dams were also seen as icons of competence and efficiency demonstrating supreme mastery over nature.

Perhaps the best example here is the construction and commissioning of the Hoover Dam from 1931 to 1936 (known at that time as the Boulder Dam), which provides flood control and electricity along the Colorado River between Arizona and Nevada. Under construction, the dam managed to attract thousands of tourists each year, and as it neared completion in 1935, more than 750,000 people visited the dam, making it as popular as tours of the Grand Canyon.[115] President Franklin D. Roosevelt turned a golden key to start the operation of the Hoover Dam in 1936, justifying the need for such dams on the grounds of creating "a new world of abundance." Pulitzer Prize winning novelist Wallace Stegner wrote in 1946 that:

> It is certainly one of the world's wonders, that sweeping cliff of concrete, those impetuous elevators, the labyrinth of tunnels, the huge power stations. Everything about the dam is marked by the immense smooth efficient beauty that seems peculiarly American.[116]

Not to be outdone, Frank Waters declared the Hoover Dam the "Great Pyramid of the American Desert" and the "Ninth Symphony of our day."[117]

Others discussed the mathematical sublimity of the dam, the sheer magnitude of the project. At the time it was the tallest building in the world, anywhere, at more than 700 feet above bedrock. Its base was as thick as two (American) football fields, its top was wide enough for three lanes of traffic. Its water intake towers were 33 stories tall, its reservoir was the largest in the world at 227 square miles, and when it was inaugurated, a double waterfall 13 feet higher than Niagara Falls was created, a feat that marked the "victory of human reason over a periodically rampaging river."[118] Indeed, when the installed electrical capacity of the dam grew from a few hundred megawatts to more than one thousand megawatts in the 1940s, the Department of the Interior proclaimed in 1946 that:

> Tomorrow the Colorado River will be utilized to the very last drop. Its water will convert thousands of additional acres of sagebrush desert to flourishing farms and beautiful homes for servicemen, industrial workers, and native farmers who seek to build permanently in the West. Its terrifying energy will be harnessed completely, to do an even bigger job in building the bulwarks for peace. Here is a job so great in its possibilities that only a nation of free people have the vision to know that it can be done and that it must be done. The Colorado River is their heritage.[119]

Such a statement situates the mastery and control of hydropower as extending well into future generations.

In addition to providing electricity, projects like the Hoover Dam created a new *social order* as they wrought benefits such as flood control, prevention of soil erosion, storage of water for irrigation, inland water transportation, and the creation of new recreational areas for fishing and boating. Such benefits were not limited to America. Lake Mead, one of the largest reservoirs in the world, was created by the Hoover Dam's blockage of the Colorado River. In addition to providing drinking water to the Las Vegas Valley Water District and cooling water for power plants, it affects the availability of water for downstream uses, including those that irrigate about a million acres of farmland in southern California's Imperial Valley, and another half million acres in northern Mexico as part of an international water treaty.[120] As another example, as part of its postwar reconstruction, Japan built five large dams on the Kitakar River designed to demonstrate "far-reaching effects as a developmental model by controlling flooding to protect communities of local people, supplying irrigation water for an increased food production, generating hydroelectricity for industrial promotion and providing tourist resources for revitalization of local areas."[121]

Lastly, dams were an integral symbol of *national pride* and even foreign policy. After erecting a series of large dams under the New Deal, one director of the Tennessee Valley Authority went so far as to claim that hydropower was "the only model on which to build our future."[122] So compelling was the mass appeal of hydroelectric dams that they became integrated with foreign policy. In 1944, Vice President Henry A. Wallace argued for building dams such as those in the

Tennessee Valley Authority throughout the world as the answer to Communist encroachment in India, Egypt, and Africa. As late as the 1960s, when the United States was losing the war in Vietnam, government leaders proposed to win the "hearts and minds" of the Vietnamese with hydroelectric dams by creating the Mekong River Authority.[123] Dams have also served as an instrument of competition in Cold War rivalries. For example, in 1955, to their great satisfaction, the Russians finished the Kuibyshev Power Plant on the Volga River. Its production of 2,300 MW outdistanced the largest American power plant, the Grand Coulee Dam, which then had a capacity of 1,974 MW.[124] We see similar geopolitical dynamics at work as lenders in China and the United States seek to "light all of Africa" by making strategic investments in the proposed Grand Inga Dam, set to be the world's largest (and most expensive) hydroelectric project in the world, at a price tag of $80 billion, if ever completed in the Democratic Republic of the Congo.[125]

Hydroelectric dams echoed the themes heard consistently in other visions of energy transitions: energy for progress, control over nature, and the provision of a new social order. It is fascinating to see that rhetoric coopted by contemporaries who oppose dams. For example, in an ongoing struggle, Brazil is attempting to build the third largest dam in the world, the Belo Monte. The Brazilian government justifies its desire by tying the plan to economic developmentalism, a popular concept in Brazil since the 1930s.[126] Mauricio Tomasquim, the president of Brazil's Energy Company (EPE) has articulated multiple reasons for Belo Monte, including a need for energy to maintain economic growth (progress), a need for more renewable energy to reduce greenhouse gas emissions (environmental progress and control), and the desire for sustainable development (social order). He notes, "Belo Monte is much more than just a mere electrical power plant but a driving force for sustainable development in the region."[127] It is interesting to note that the opposition frequently picks up the same visions. For example, Candido Grzybowski of the Brazilian Institute of Social and Economic Analysis rejects the desire for the Belo Monte dam but concludes her objections by observing, "The debate on the Hydroelectric of Belo Monte is, above all, a debate on the Brazil that we want—sustainable, in solidarity and democratic—in which power and economy relocate, nearer to the citizen and are controlled by him/her."[128]

In sum, hydroelectric dams during this era were able to influence the social imagination for their ability to bring natural forces under human control and democratize the fruits of economic development. Perhaps more than any other type of infrastructure, such dams were able to represent the progress of humanity from a life ruled by natural forces and tradition to one where nature is ruled by engineering and tradition by rationality and modernity.[129]

Nuclear reactors: 1910s to present

Like hydroelectric dams, nuclear reactors also intersected with visions about mastery and control over nature, utopian social orders, and symbols of national pride. Even before the dawn of the nuclear era, proponents depicted nuclear power as an alluring option to create energy at low to no cost. In 1914, H.G. Wells' *The World Set Free* formulated a future in which the atom was used for peaceful purposes, spawning a cultural renaissance. When Frederick Soddy was awarded the Nobel Prize in 1921 for his work on radioactive isotopes, he commented that if humanity could liberate the energy locked in the heart of an atom, it would have an "inexhaustible supply of power to transform society."[130]

Nuclear power strongly associated with *mastery over nature* coupled with new *social orders*. Early advocates promised a future of electricity too cheap to meter, an age of peace and plenty without high prices and shortages where atomic energy provided the power needed to desalinate water for the thirsty, irrigate deserts for the hungry, and fuel interstellar travel deep into outer space (see Chapter Two's section on "Utopianism and Fantasy in Energy Systems" for more on similar visions about a nuclear future). Other exciting opportunities included atomic golf balls that could always be found, a nuclear powered car designed by the Ford Motor Company, and a nuclear powered airplane, which the United States Federal Government spent $1.5 billion researching between 1946 and 1961.[131]

One explanation for the attractiveness of nuclear energy could be its association with *national visions of progress*. While these visions vary by country and over time, John Byrne and Steve Hoffmann propose that the single most consistent predictor of whether a society will embrace nuclear energy is their ability to think in the "future tense."[132] That is, planners and promoters will become enthralled by the possible benefits of nuclear energy in the *future* and are willing to accept the costs in the *present* to realize them. Put another way, they will overestimate the advantages from nuclear energy and discount its future costs in the absence of knowledge about current economic or technical compatibility; reality of present risks and costs are discounted by the unrealized (and perhaps unrealizable) possibilities of future gain. Although these psychological benefits are intangible, they are often believed to be real. A historical examination of the genesis of nuclear power programs in numerous countries reveals that, in each case, optimism in the technology and an overarching vision of what nuclear energy could deliver in the future played a role in trumping concerns about present costs.[133] Table 1.1 offers a concise summary of these views.

In China, nuclear power was seen as instrumental in overcoming energy supply deficits, improving Chinese economic competitiveness, "catching up" with Taiwan and other industrialized countries, and enhancing national prestige. Chinese officials even toyed with the idea of exporting both nuclear technology and electricity to the rest of Asia and built one facility, the Yibin Fuel Component Factory in Sichuan, to manufacture prefabricated components of nuclear power plants for export. They sold one set of components to Pakistan in 1989 and planned to earn billions of dollars of foreign exchange exporting similar packages to Africa and the rest of the developing world.[135]

Left in the devastation by the German occupation and fighting of 1944 to 1945, French technical and scientific experts linked nuclear power to French

TABLE 1.1 Nuclear power visions across nine national programs

Country	Period	Vision
China	1953 to 1992	"Catching up" with other industrialized countries (including Taiwan) and creating lucrative opportunities for Chinese exports and economic leadership
France	1945 to 1970	Recovering from World War II and revitalizing the national economy through high technology "national champions" that would legitimate France as a vital superpower
India	1945 to 1980	Creating a prosperous and technologically sophisticated Indian society where social problems (such as hunger and poverty) would be eliminated
Japan	1955 to 1990	Using technological prowess and nuclear energy to rebuild the national economy and offset the risks of energy shortages and dependence on energy imports
Soviet Union	1954 to 1986	Validating the communist system and Soviet approach to science and achieving a utopian future without scarcities of water, food, heat, and energy
United States	1942 to 1979	Harnessing the power of the atom for peaceful purposes, legitimatizing the Manhattan Project, and creating a future where electricity was "too cheap to meter"
Spain	1951 to 1980	Revitalizing the Spanish economy after World War II and participating in "important" scientific research involving fission
South Korea	1956 to 2018	Signaling to the rest of the world that South Korea was moving from a technologically backward to a technologically advanced nation, and from militarily weak to strong
Canada	1943 to 2018	Cultivating global demand for Canadian uranium and creating a lucrative export market for Canadian reactors

Source: Author, modified from Sovacool, B. K., and S. V. Valentine 2012[134]

radiance and identity.[136] Nuclear energy was central to this campaign of French economic modernization, and research, development, and construction were dominated by the government. Nuclear reactors offered the chance for French planners to rebuild infrastructure, promote industry, and augment political influence simultaneously. After the creation and demonstration of the atomic bomb, historian Gabrielle Hecht notes that for France "nuclear technology became a quintessential symbol of modernity and national power."[137] James M. Jasper provocatively described nuclear power development since then in France as consisting of three groups—"*gods*" (governments), "*titans*" (large industries and utilities), and "*mortals*" (the general public). Jasper suggested that success of the French program stemmed from the inability of the "mortals" to interfere in the workings of the gods and titans.[138]

In India, planners conceived of the national nuclear program as key to confirming the country's place in the modern era and intersected with the widely held belief that energy abundance underpinned social progress. Prime Minister Jawaharlal Nehru argued in 1948 that India had failed to capitalize on the first Industrial Revolution due to lack of technical skill, and believed that success in the ongoing second Industrial Revolution was predicated on engineering prowess, typified by nuclear power. Later in the 1970s, Prime Minister Indira Gandhi reiterated Nehru's position that nuclear power was as an essential technology for rescuing developing economies such as India's from "poverty and ignorance." She was convinced that a bold display of scientific and technological might could impress the populace enough to win her reelection.[139]

In Japan, the promise of generating cheap energy through applied nuclear technology meshed perfectly with government aspirations to enhance the international competitiveness of industry. National planners came to see nuclear technology as an important export product, a tool to not only free the nation from energy dependence but to also extend its economic reach into the Pacific and the world at large. The sheer lack of indigenous energy resources justified a massive expansion of the nuclear program, including commitment to plutonium fueled fast breeder reactors. Japanese officials believed that a greater national risk was posed by dependence on imported energy than by a network of nuclear power plants.[140]

Nuclear energy was quickly attached to the infallibility of Soviet science and technology as well as the idea of a progressive communist regime free from energy shortages and wants. Atomic energy came to be not only a source of electricity supply for government planners but also a pathway towards developing breeder reactors that would meet all of the country's energy needs, a first step towards perfecting nuclear powered engines for aircraft and automobiles, a system for producing radiation to preserve food, a source of knowledge about nuclear technology that could help the Soviet Union build advanced weapons, and a mechanism of political control where planners dispersed nuclear reactors to the republics to strengthen ties and political adherence. It also went hand-in-hand with an agenda to convert an agrarian and peasant society into a "well-oiled machine of workers" tirelessly committed to communism.[141,142]

In the United States, nuclear energy promotion reinforced national values and ideas about technology and nature. Anthropologist Gary Downey has argued that advanced technology has always been correlated with progress in the United States, initially used to distinguish the American colonies from their English counterparts.[143] Nuclear energy was seen as politically necessary to avoid the risks of communism, and key to a postwar identity shaped in defiance to Marxism. Military planners believed that demonstrating the civilian applications of the atom would also affirm the American system of private enterprise and the expertise of its scientists, while increasing personal living standards and defending a democratic lifestyle.

Spain pursued a path of nuclear power partly because of their technocratic government, imperialist ambitions, utopian thinking, and Cold War relationships. As a consequence of its dictatorship, and its collaboration with the Third Reich during World War II, Spain was excluded from international forums until 1955, and it did not receive economic aid under the Marshall Plan. Impoverished by war, Spanish planners saw nuclear energy as an inexhaustible source of energy necessary to power Spain's national reconstruction, development and industrialization.[144]

South Korean leaders also embraced nuclear power for visionary and symbolic reasons. Nuclear power served as a signal to the rest of the world that South Korea was moving from a technologically backward to a technologically advanced nation, and from militarily weak to strong. Put another way, government elites saw nuclear power as central to lifting South Korea out of impoverishment after civil war.[145] Their vision also centered on designing indigenous reactors that would both enhance energy security and increase economic competitiveness. As Korean scholar Tay Joon Lee has written, "Korean nuclear power development has been inextricably linked to the enhancement of energy security for national economic development and industrialization."[146]

Finally, Canada's nuclear program has been intertwined with a vision of capacity building in terms of nurturing nuclear technology and the development of the CANDU reactor, which has been recognized by the Canadian Engineering Centennial Board as one of the 10 most outstanding Canadian engineering accomplishments in the twentieth century.[147] It was also hoped the nuclear program would transform Canada into a key producer of medical isotopes and a leading center for nuclear medicine research. In short, although Canada's nuclear power program was propagated in part by technocratic ambitions, the program in turn played a symbolic role in further fortifying technocratic ideology in Canada.

Consider that during these early years of the nuclear energy industry across these countries, vendors were able to sell the technology not because it was necessarily superior to alternative technologies at the time but because it coalesced nicely with preconceived notions of progress and modernity, and was coupled with statements from credible sources such as industry leaders and government spokespersons.

Discursive themes and utopian dynamics

The future, of course, did not turn out entirely the way that the advocates of these four visions hoped for. Perhaps their past thinking surprises us as much as they would have been surprised about what has become of the future which they dreamed of. Nonetheless, an analysis of these four energy transitions suggests discursive themes that recur with frequency, such as mastery and control over nature; utopian social and technological order; and symbols of national pride. Not only do these themes repetitiously occur, once they are established, they lend themselves for use in other arenas. For example, hydroelectric power not only demonstrated that dams became a symbol of American pride but also that once they were such symbols, policymakers could advocate the construction of dams in other parts of the world as tools for achieving foreign policy objectives. What is most important about these themes, however, is that rhetorical visions tend to exist independent of the ability of the technology to produce energy, which is remarkable given that I am discussing energy transitions.

Mastery and control over nature

Although some human societies still attempt to live within nature, the desire to control nature is at the core of human advancement. What constitutes mastery and how control is measured may change, but the theme recurs in three of the four transitions and, if examined from a slightly different perspective, emerges in transition to gasoline powered automobiles. Steam engines provided control over energy sources as they were more predictable than others, and more control over labor as the engines never tired. In a sense of environmental control completely alien to today's planners, the pollution produced by the engines even symbolized prosperity through nature. For hydroelectric dams, the control over nature is even more evident. Building a dam is, by its very nature, the exercise of controlling the flow of rivers. Similarly, the proposed ability to turn a handful of snow into power for a city reflects the nuclear claim of mastery over nature. Each suggests taking something as natural as water, atoms, or fossilized hydrocarbons, and converting it into the energy needed to power civilization.

The gasoline automobile transition does not demonstrate the same control over nature, probably because it is not a source of energy, and instead is a device that uses energy. While there is an argument that crude oil is also "natural," the fact that it is buried and therefore out of the public's sight, makes the control over nature much less obvious. However, the advertisement of automobiles as a way to see nature may be a circuitous route to a similar theme. That coupled with the nature of taking control over one's own transportation suggests that similar motives were in play, even if the more clearly evident idea of enhancing mastery over nature is missing.

Utopian social and technological order

One of the most powerful of the rhetorical themes surrounding energy transitions is the ability to radically transform social and technological order. Steam engines transformed the workplace and provided a powerful rationale for the advent of the service industry. Gasoline powered cars changed the transportation patterns of much of the world and set the stage for the development of the modern urban landscape. Hydroelectric dams dictated agricultural and economic development for entire regions, and as the Belo Monte development project demonstrates, create debates over who should be empowered by those changes. The atomic age suggested a complete transformation of almost all elements of human order, from an energy source to replace all others, to radioactive makeup and golf balls.

The importance of this theme lies in its ability to project future realities in which the new energy source dominates. Given that we use energy to *do things*, one of the key questions in any energy transition becomes, "what *new* things can we do?" It is likely that support for any particular energy transition increases as the vision for what we can do with that energy becomes more exciting. This may also explain why, despite its successes, public enthusiasm for conservation, energy efficiency, demand response, or demand side management efforts often remains relatively low. After all, conservation is generally perceived as finding ways to do less than what we currently do. While a worthwhile goal, it suffers in comparisons to exciting new energy forms which introduce no such constraints.

Symbols of national pride

Symbols of national pride are another prominent theme. The steam engine represented strength and elegance as well as scientific prowess. The construction and marketing of automobiles is a key segment of all major economies. The quality of car is a status symbol, demonstrating both a national and personal pride. For example, China overtook the United States in 2009 as the largest car market in the world, with car brands tied to specific levels of economic status (BMW 5 series for middle management; Audi 6 for government officials and business owners; Mercedes Benz E series for old money, etc.).[148] The national pride in dams was evident in the desire to use dam building as a way to expand U.S. influence during the Cold War. While nuclear power has ebbed and flowed based on a variety of factors, including well publicized accidents, the national pride arguments are evident in the visions associated with nuclear power in nine national programs.

Chapters to come

These four waves of historical excitement over steam engines, cars, hydroelectric dams, and nuclear reactors give rise to some pressing contemporary questions. They signify that the co-production of technological innovation and social ordering fundamentally begins with processes of envisaging the future. But how does language, visions, and discourses shape the politics of new and emerging low-carbon technologies and systems? Is there a common rhetorical structure to future visions about energy, climate sustainability, or mobility? Are the visions and expectations surrounding energy transitions different this time around? What are the dynamics of expectations across different types of publics, notably experts or ordinary citizens? Lastly, how might expectations and visions conflict or hold radically different underlying values or assumptions?

The chapters to come explore these themes and questions in more intricate detail. They suggest that visionary discourses have such provocative appeal because they satisfy an underlying social need, despite their distinctive elements. In recent decades it has become clear that the current economic paradigm has run up against environmental limits. For the system to continue in roughly a business-as-usual fashion, it needs to find sources of energy that emit low quantities of carbon, are risk free and virtually inexhaustible. In addition, as natural resources are becoming scarce, the possibility of exploring outer space seems the next obvious source of material abundance. This means that energy technologies, and indeed all types of technology, may be chosen not entirely because of their utility—the ability to produce kilowatt hours, facilitate mobility, or generate desalinated water—but because they capture imaginations, confirm an ideology, or fit with a particular blueprint about the future.[149]

In proceeding to chart, analyze, construct, and even deconstruct energy visions, the chapters ahead themselves in some ways tell stories, and these stories are not always neat or linear. But even in their tensions and contradictions, they still accurately reflect society's own ambiguities in grapping and engaging with low-carbon systems and their affiliated futures. The energy systems of tomorrow will be processed, debated, and imagined in a tension-filled and contested arena of discursive politics. Whose, and which, vision ultimately wins and becomes dominant will be entirely dependent on the processes by which these visions are (and have been) socially and politically constructed. Hence the necessity of critically examining them in the pages to come.

Notes

1 Polak, F. L. 1973. The Image of the Future (San Francisco, CA: Jossey-Bass).
2 Bertaux, P. 1968. The Future of Man. In Ewald, W. R. (Ed.): Environment and Change: The Next Fifty Years (London: Indiana University Press), pp. 13–20.
3 Tenner, E. 2006, Winter. The Future Is a Foreign Country. The Wilson Quarterly 30 (1): 62–66.
4 Boulding, E., and K. E. Boulding. 1995. The Future: Images and Processes (London: Sage Publications).
5 Boulding, K. 1977. Determinants of Energy Strategies. In Kiefer, I. (Ed.): Future Strategies for Energy Development: A Question of Scale (Knoxville, TN: Oak Ridge National Laboratory and U.S. Department of Energy), p. 17.
6 Fujimura, J. (2003). Future Imaginaries: Genome Scientists as Socio-Cultural Entrepreneurs. In Goodman, A. H., Heath, D., and Susan Lindee, M. (Eds.): Genetic Nature/Culture: Anthropology and Science beyond the Two-Culture Divide (Berkeley, CA: University of California Press), pp. 176–199.
7 Van Lente, H. 1993. Promising Technologies: The Dynamics of Expectations in Technological Developments (Enschede, NL: Twente University).
8 Jasanoff, S., and S.-H. Kim. 2009. Containing the Atom: Sociotechnical Imaginaries and Nuclear Power in the United States and South Korea. Minerva 47 (2): 119–146.
9 Borup, M., N. Brown., K. Konrad., and H. V. Lente. 2006. The Sociology of Expectations in Science and Technology. Technology Analysis & Strategic Management 18 (3-4): 285–29
10 Selin, C. 2008. The Sociology of the Future: Tracing Stories of Technology and Time. Sociology Compass 2 (6): 1878–1895.
11 Nightingale, P. 1998. A Cognitive Model of Innovation. Research Policy 27 (7): 689–709.
12 Sanz-Menendez, L., and C. Cabello. 2000. Expectations and Learning as Principles for Shaping the Future. In Brown, N., Rappert, B., and Webster, A. (Eds.): Contested Futures: A Sociology of Prospective Techno-Science (Aldershot; Burlington, VT: Ashgate), p. 231.
13 Mason, A. 2006. Images of the Energy Future. Environmental Research Letters 1 (1): 014002.

14 Hecht, G. 1998. The Radiance of France: Nuclear Power and National Identity after World War II. Inside Technology (Cambridge, MA: MIT Press).

15 McDougall, W. A. 2008. The Heavens and the Earth: A Political History of the Space Age (New York: Basic Books).

16 Neufeld, M. J. 1995. The Rocket and the Reich: Peenemünde and the Coming of the Ballistic Missile Era (Cambridge, MA: Harvard University Press).

17 Schelhas, J., S. Hitchner, and J. P. Brosius. 2018, January. Envisioning and Implementing Wood-Based Bioenergy Systems in the Southern United States: Imaginaries in Everyday Talk. Energy Research & Social Science 35: 182–192.

18 Kuchler, M. 2014. Sweet Dreams (Are Made of Cellulose): Sociotechnical Imaginaries of Second-Generation Bioenergy in the Global Debate. Ecological Economics 107: 431–437.

19 Fatimah, Y. A. 2015. Fantasy, Values, and Identity in Biofuel Innovation: Examining the Promise of Jatropha for Indonesia. Energy Research & Social Science 7: 108–116.

20 Levidow, L., and T. Papaioannou. 2013. State Imaginaries of the Public Good: Shaping UK Innovation Priorities for Bioenergy. Environmental Science & Policy 30: 36–49.

21 Mittlefehldt, S. 2016, April. Seeing Forests as Fuel: How Conflicting Narratives Have Shaped Woody Biomass Energy Development in the United States since the 1970s. Energy Research & Social Science 14: 13–21.

22 Giurca, A. In press. Unpacking the Network Discourse: Actors and Storylines in Germany's Wood-Based Bioeconomy. Forest Policy and Economics. Corrected proof, Available online 29 May 2018.

23 Sherren, K. et al. 2016, April. Learning (Or Living) to Love the Landscapes of Hydroelectricity in Canada: Eliciting Local Perspectives on the Mactaquac Dam via Headpond Boat Tours. Energy Research & Social Science 14: 102–110.

24 Ballo, I. F. 2015. Imagining Energy Futures: Sociotechnical Imaginaries of the Future Smart Grid in Norway. Energy Research & Social Science 9: 9–20.

25 Hommes, L., and R. Boelens. 2018. From Natural Flow to "Working River": Hydropower Development, Modernity and Socio-Territorial Transformations in Lima's Rímac Watershed. Journal of Historical Geography.

26 Dukpa, R. D., D. Joshi, and R. Boelens. 2018. Hydropower Development and the Meaning of Place. Multi-Ethnic Hydropower Struggles in Sikkim, India. Geoforum 89: 60–72.

27 Korsnes, M. 2016, June. Ambition and Ambiguity: Expectations and Imaginaries Developing Offshore Wind in China. Technological Forecasting and Social Change 107: 50–58.

28 Karlsen, A. 2018, January. Framing Industrialization of the Offshore Wind Value Chain—A Discourse Approach to an Event. Geoforum 88: 148–156.

29 Mohan, A., and K. Top. 2018, October. India's Energy Future: Contested Narratives of Change. Energy Research & Social Science 44: 75–82.

30 Jepson, W., C. Brannstrom, and N. Persons. 2012, June. "We Don't Take the Pledge": Environmentality and Environmental Skepticism at the Epicenter of US Wind Energy Development. Geoforum 43 (4): 851–863.

31 Toonen, H. M., and H. J. Lindeboom. 2015, February. Dark Green Electricity Comes from the Sea: Capitalizing on Ecological Merits of Offshore Wind Power? Renewable and Sustainable Energy Reviews 42: 1023–1033.

32 Cloke, J., A. Mohr, and E. Brown. 2017, September. Imagining Renewable Energy: Towards a Social Energy Systems Approach to Community Renewable Energy Projects in the Global South. Energy Research & Social Science 31: 263–272.

33 Simmet, H. R. 2018, June. "Lighting a Dark Continent": Imaginaries of Energy Transition in Senegal. Energy Research & Social Science 40: 71–81.

34 Rosenbloom, D. et al. 2016, July. Framing the Sun: A Discursive Approach to Understanding Multi-Dimensional Interactions within Socio-Technical Transitions

through the Case of Solar Electricity in Ontario, Canada. Research Policy 45 (6): 1275–1290.

35 Curran, G. 2012, February. Contested Energy Futures: Shaping Renewable Energy Narratives in Australia. Global Environmental Change 22 (1): 236–244.

36 Phillips, M., and J. Dickie. 2014, April. Narratives of Transition/Non-Transition Towards Low Carbon Futures within English Rural Communities. Journal of Rural Studies 34: 79–95.

37 Bozeman, B. 1977, September–October. Epistemology and Future Studies: How Do We Know What We Can't Know? Public Administration Review 37 (5): 544–549.

38 Freund, J. 2015. Rev Billy vs. the Market: A Sane Man in a World of Omnipotent Fantasies. Journal of Marketing Management 31: 1529–1551.

39 Freund, J., and E. S. Jacobi. 2015. "Mystify Me: Coke, Terror and the Symbolic Immortality Boost", in Special Section: "Marketing as Mystification". Marketing Theory 16: 417–421.

40 Noppers, E. H. et al. 2014. The Adoption of Sustainable Innovations: Driven by Symbolic and Environmental Motives. Global Environmental Change 25: 52–62.

41 Miller, C. A. et al. 2015, June. Narrative Futures and the Governance of Energy Transitions. Futures 70: 65–74.

42 Bormann, E. G. 1982. Fantasy and Rhetorical Vision: Ten Years Later. Quarterly Journal of Speech 68: 288–305.

43 Bakker, S., H. Van Lente, and M. Meeus. 2011. Arenas of Expectations for Hydrogen Technologies. Technological Forecasting and Social Change 78 (1): 152–162.

44 Selin, C. 2007. Expectations and the Emergence of Nanotechnology. Science, Technology & Human Values 32 (2): 196–220.

45 Berkout, F. 2006, July–September. Normative Expectations in Systems Innovation. Technology Analysis & Strategic Management 18 (3/4): 299–311.

46 Brown, N., B. Rappert, and A. Webster. 2000. Introducing Contested Futures: From Looking into the Future to Looking at the Future. In Brown, N., Rappert, B., and Webster, A. (Eds.): Contested Futures: A Sociology of Prospective Techno-Science (Aldershot; Burlington, VT: Ashgate), pp. 3–20.

47 Cozen, B., D. Endres, T. R. Peterson, C. Horton, and J. T. Barnett. 2018. Energy Communication: Theory and Praxis Towards a Sustainable Energy Future. Environmental Communication 12 (3): 289–294.

48 Sovacool, B. K., and B. Brossmann. 2013, June. Fantastic Futures and Three American Energy Transitions. Science as Culture 22 (2): 204–212.

49 Merton, R. K. 1973. The Sociology of Science (Chicago, IL: University of Chicago Press).

50 Kuhn, T. S. 1962. The Structure of Scientific Revolutions (Chicago, IL: University of Chicago Press).

51 Kuhn, T. S. 1977. The Essential Tension: Selected Studies in Scientific Tradition and Change (Chicago, IL: University of Chicago Press).

52 Price, D. D. S., and D. Beaver. 1966. Collaboration in an Invisible College. American Psychologist 21: 1011–1018.

53 Ceccarelli, L., R. Doyle, and J. Selzer. 1996. Introduction to the Special Issue on Rhetoric of Science. Rhetoric Society Quarterly 26 (4): 7–12.

54 Ceccarelli, L. 2004. Rhetoric of Science and Technology. In Mitchem, C. (Ed.): Encyclopedia of Science, Technology, and Ethics, Vol. 3: L-R (Detroit, MI: Macmillan Reference), pp. 1625–1629.

55 Ceccarelli, L. 2005. A Hard Look at Ourselves: A Reception Study of Rhetoric of Science. Technical Communication Quarterly 14 (3): 257–265.

56 Taylor, C. A. 1996. Defining Science: A Rhetoric of Demarcation (Madison, WI: University of Wisconsin Press).

57 Porter, T. 1995. Trust in Numbers: The Pursuit of Objectivity in Science and Public Life (Princeton, NJ: Princeton University Press).

58 Beck, U. 1992. Risk Society: Towards a New Modernity (London: Sage Publications).

59 Gross, A. G. 2006. Starring the Text: The Place of Rhetoric in Science Studies (Carbondale, IL: Southern Illinois University Press).

60 Barry, J., G. Ellis, and C. Robinson. 2008, May. Cool Rationalities and Hot Air: A Rhetorical Approach to Understanding Debates on Renewable Energy. Global Environmental Politics 8 (2): 67–98.

61 Latour, B. 1987. Science in Action: How to Follow Scientists and Engineers through Society (Cambridge, MA: Harvard University Press, 1987).

62 Collins, H. M. 1985. Changing Order: Replication and Induction in Scientific Practice (Chicago, IL: University of Chicago Press).

63 Eisenstein, E. 1983. The Printing Revolution in Early Modern Europe (Cambridge: Cambridge University Press).

64 McLuhan, M. 1964. Understanding Media: The Extensions of Man (New York: Routledge).

65 Postman, N. 1986. Amusing Ourselves to Death: Public Discourse in the Age of Show Business (New York: Penguin Press).

66 Rifkin, J. 2000. The Age of Access: The New Culture of Hypercapitalism Where All of Life Is a Paid-For Experience (New York: Penguin Putnam).

67 Bijker, W. 1993. Do Not Despair: There Is Life after Constructivism. Science, Technology, & Human Values 18 (1): 113–138.

68 Olson, R. L. 1973. Sustainability as a Social Vision. Journal of Social Issues 51: 15–35.

69 Segal, H. P. 1994. Future Imperfect: The Mixed Blessings of Technology in America (Boston, MA: University of Massachusetts).

70 Segal, H. P. 2005. Technological Utopianism in American Culture (Syracuse, NY: Syracuse University Press).

71 Berkhout, F. 2006. Normative Expectations in Systems Innovation. Technology Analysis & Strategic Management 18: 299–311.

72 Geels, F. W., and W. A. Smit. 2000. Failed Technology Futures: Pitfalls and Lessons from a Historical Survey. Futures 32: 867–885.

73 Bazerman, C. 1999. The Languages of Edison's Light (Cambridge, MA: MIT Press).

74 Burke, K. 1950. A Rhetoric of Motives (Berkeley, CA: University of California Press).

75 Burke, K. 1945. A Grammar of Motives (New York: Prentice-Hall).

76 Thornhill, J. 2018, February 7. Forget the Bravado—Elon Musk's SpaceX Achievement Is Stunning. Financial Times.

77 Burke, K. 1966. Language as Symbolic Action: Essays on Life, Literature, and Method (Berkeley, CA: University of California Press).

78 Bormann, E. G. 1972. Fantasy and Rhetorical Vision: The Rhetorical Criticism of Social Reality. Quarterly Journal of Speech 58: 396–407.

79 Brown, N., and M. Michael. 2003. A Sociology of Expectations: Retrospecting Prospects and Prospecting Retrospects. Technology Analysis & Strategic Management 15 (1): 3–18.

80 Sovacool and Brossmann (2013: 204–212).

81 Sovacool, B. K., and B. Brossmann. 2014, September. The Rhetorical Fantasy of Energy Transitions: Implications for Energy Policy and Analysis. Technology Analysis & Strategic Management 26 (7): 837–854.

82 Patterson, W. 2007. Keeping the Lights On: Towards Sustainable Electricity (London: Earthscan).

83 Ferguson, E. S. 1980. The Nature of the Steam Engine. In Kranzberg, M., Hall, T. A., and Scheiber, J. L. (Eds.): Energy and the Way We Live (San Francisco, CA: Boyd & Fraser Publishing), pp. 79–83.

84 Science June 29, 1899, vol X1, no. 282, p. 303.

85 Basalla, G. 1982. Some Persistent Energy Myths. In Daniels, G. H. and Rose, M. H. (Eds.): Energy and Transport: Historical Perspectives on Policy Issues (London: Sage Publications), pp. 27–29.

86 Centennial Corliss Engine Groups. 2002. The Corliss Steam Engine of 1876. (Philadephia, Pennsylvania: Centennial Corliss Engine Groups).

87 Wang, H.-C. 2009, March. Discovering Steam Power in China. Technology and Culture 51 (10): 31–54.

88 Nye, D. E. 1999. Consuming Power: A Social History of American Energies (Cambridge, MA: MIT Press).

89 Tenner, E. 1997. Why Things Bite Back: Technology and the Revenge of Unintended Consequences (New York: Alfred A. Knopf).

90 Nye, D. E. 1994. American Technological Sublime (Cambridge, MA: MIT Press).

91 Barbour, I., H. Brooks, S. Lakoff, and J. Opie. 1982. Energy and Abundance: Advance and Retreat. In Energy and American Values (Westport, CT: Praegar), pp. 24–43.

92 Ibid., 26.

93 Ibid., 26.

94 Ibid., 29.

95 Sovacool and Brossmann (2014: 837–854).

96 Sovacool, B. K. 2009. Early Modes of Transport in the United States: Lessons for Modern Energy Policymakers. Policy & Society 27: 411–427.

97 Belasco, R. J. 1982. Cars versus Trains: 1980 and 1910. In Daniels, G. H. and Rose, M. H. (Eds.): Energy and Transport: Historical Perspectives on Policy Issues (London: Sage), pp. 39–53.

98 Kirsch, D. A. 2000. The Electric Vehicle and the Burden of History (London: Rutgers University Press).

99 Mom, G. P. A., and D. A. Kirsch. 2001. Technologies in Tension: Horses, Electric Trucks, and the Motorization of American Cities, 1900 to 1925. Technology & Culture 42: 489–518.

100 Sagoff, M. 2008. The Economy of the Earth: Philosophy, Law, and the Environment (Cambridge: Cambridge University Press).

101 Glaab, C. N., and A. T. Brown. 1967. A History of Urban America (New York: MacMillan Company).

102 Mitchell, W. J., C. E. Borroni-Bird, and L. D. Burns. 2010. Reinventing the Automobile: Personal Urban Mobility for the 21st Century (Cambridge, MA: MIT Press), pp. 12–13.

103 International Organization of Motor Vehicle Manufacturers, Economic Impact. 2007. http://oica.net/category/economic-contributions/.

104 International Organization of Motor Vehicle Manufacturers, World Motor Vehicle Production and OICA. 2018. 2017 Production Statistics. www.oica.net/category/production-statistics/2017-statistics/.

105 U.S. Department of Energy. 2010. Fact #617: April 5, 2010 Changes in Vehicles per Capita around the World. www1.eere.energy.gov/vehiclesandfuels/facts/2010_fotw617.html.

106 Ibid.

107 Sperling, D., and D. Gordon. 2010. Two Billion Cars: Driving toward Sustainability (New York: Oxford University Press).

108 Nye (1999).

109 Basalla, G. 1982. Some Persistent Energy Myths. In Daniels, G. H. and Rose, M. H. (Eds.): Energy and Transport: Historical Perspectives on Policy Issues (London: Sage Publications), pp. 27–29.

110 Melosi, M. V. 1982. Energy Transitions in the Nineteenth-Century Economy. In Daniels, G. H. and Rose, M. H. (Eds.): Energy and Transport: Historical Perspectives on Policy Issues (London: Sage), pp. 55–69.

111 Nye, D. E. 2003. America as second creation: Technology and narratives of new beginnings. (Cambridge, MA: MIT Press).

112 Nye (1994: 134).

113 Quoted in Ibid., 135.

114 Ibid., 137.

115 Ibid., 134.

116 Bocking, S. 2009, February 15. Dams as Development. Globalization Monitor, p. 1.

117 Nye (1994).

118 Nye (2003: 140).

119 Nye (2003: 246).

120 Sovacool, B. K., and K. E. Sovacool. 2009, July. Identifying Future Electricity Water Tradeoffs in the United States. Energy Policy 37 (7): 2763–2773.

121 Berga, J. M. et al. 2006. Dams and Reservoirs, Societies and Environment in the 21st Century (New York: Taylor and Francis 2006), p. 34.

122 Feiss, C. 1968. Taking Stock: A Resume of Planning Accomplishments in the United States. In Ewald, W. R. (Ed.): Environment and Change: The Next Fifty Years (London: Indiana University Press), pp. 214–236.

123 Basalla (1982).

124 Berga et al. (2006).

125 Green, N., B. K. Sovacool, and K. Hancock. 2015, April. Grand Designs: Assessing the African Energy Security Implications of the Grand Inga Dam. African Studies Review 58 (1): 133–158.

126 Nunes, R. 2011, February 15. What's behind the Belo Monte Dam? The Guardian. www.guardian.co.uk/commentisfree/cifamerica/2011/feb/15/brazil-energy.

127 Tolmasquim, M. 2011, February 1. Transcript: Conference Call with Mauricio Tolmasquim on Belo Monte Dam Project Teleconference Held by the President of Brazil's Energy Company (EPE) to International Media. www.brasil.gov.br/para/press/conferences/february-1/transcript-conference-call-with-mauricio-tolmasquim-on-belo-monte-dam-project/br_model1?set_language=en.

128 Grzybowski, C. 2012. Hydroelectric of Belo Monte: A Question of Democracy (Brazilian Institute of Social and Economic Analysis). www.cimi.org.br/?system=news&action=read&id=4447&eid=402.

129 Khagram, S. 2005. Beyond Temples and Tombs: Towards Effective Governance for Sustainable Development through the World Commission on Dams (Cambridge, MA: Center for International Development and Hauser Center for Non-Profit Organizations, The John F Kennedy School of Government Harvard University).

130 Basalla (1982).

131 Duncan, O. D. 1978, September. Sociologists Should Reconsider Nuclear Energy. Social Forces 57 (1): 1–22.

132 Byrne, J., and S. M. Hoffman. 1996. The Ideology of Progress and the Globalisation of Nuclear Power. In Byrne, J. and Hoffman, S. M. (Eds.): Governing the Atom: The Politics of Risk (London: Transaction Publishers), pp. 11–46.

133 Sovacool, B. K., and S. V. Valentine. 2012. The National Politics of Nuclear Power: Economics, Security, and Governance (London: Routledge).

134 Ibid.

135 Gallagher, M. G. 1990. Nuclear Power and Mainland China's Energy Future. Issues and Studies 26 (12): 100–120.

136 Scheinman, L. 1965. Atomic Energy Policy in France under the Fourth Republic (Princeton, NJ: Princeton University Press).

137 Hecht (1998).

138 Jasper, J. M. 1992. Gods, Titans and Mortals: Patterns of State Involvement in Nuclear Development. Energy Policy 20 (7): 653–659.

139 Sovacool and Valentine (2012).

140 Sovacool and Valentine (2012).

141 Josephson, P. R. 1995. "Projects of the Century" in Soviet History: Large-Scale Technologies from Lenin to Gorbachev. Technology & Culture 36 (3): 519–559.

142 Josephson, P. R. 1999. Red Atom: Russia's Nuclear Power Program from Stalin to Today (New York: W. H. Freeman).

143 Downey, G. L. 1986. Risk in Culture: The American Conflict over Nuclear Power. Cultural Anthropology 1 (4): 388–412.

144 Puig, A., and I. Presas. 2005. Science on the Periphery: The Spanish Reception of Nuclear Energy. Minerva 43: 197–218.

145 Byrne and Hoffman (1996: 11–46).

146 Lee, T. J. 2004. Technological Change of Nuclear Fuel Cycle in Korea. Progress in Nuclear Energy 45 (1): 87–104.

147 Sovacool and Valentine (2012).

148 Barton, R. L. 2011, June 11. Chinese Snatch up Status Symbol Cars China in Focus. http://chinainfocus.net/?p=1348.

149 Elzen, B., F. W. Geels, and K. Green (Eds.). 2004. System Innovation and the Transition to Sustainability: Theory, Evidence and Policy (Cheltenham, UK: Edward Elgar).

2

TECHNOLOGICAL UTOPIANISM

Small modular reactors and the physics community

When the Pulitzer Prize winning novelist Wallace Stenger visited the Hoover Dam in 1946, he wrote in his diary that "it is certainly one of the world's wonders, that sweeping cliff of concrete, those impetuous elevators, the labyrinth of tunnels, the huge power stations."[1] U.S. President Franklin D. Roosevelt's reaction was even more concise: "I came, I saw, and I was conquered."[2] Their remarks reveal how some large-scale technologies, and forms of energy infrastructure, can provoke an almost religious and collective feeling of sublimity.[3] They can inspire much like the cathedrals and temples of old did our ancestors, capturing public imagination (and at times devotion).[4]

In this chapter, I demonstrate (extending previous work[5]) that technical scientists and engineers are perhaps as zealous in their promotion of the concept of small modular nuclear reactors, or SMRs, as novelists and presidents were in endorsing hydroelectric dams more than half a century ago. The focus on SMRs follows in the coattails of the much-heralded notion of a renaissance in nuclear power, punctuated by the accidents at Fukushima. After introducing readers to various SMR designs and concepts, as well as the literature on the topic of utopianism and fantasy, the chapter identifies five distinct rhetorical visions of SMRs focused on risk-free energy, self-energization, water security, environmental nirvana, and space exploration. The importance of identifying these SMR visions is threefold.

First, understanding the dynamics constraining or accelerating nuclear power reactors, as well as the epistemological assumptions underpinning the expansion of the industry, is essential to properly weighing its costs, benefits, and future role. Even after the Fukushima nuclear accident in 2011, many analysts have argued that the world remains on the cusp of a "nuclear renaissance," "nuclear resuscitation," and a "second nuclear era."[6,7,8,9] Other studies proclaim that SMRs will be the "savior of the nuclear renaissance"[10] and represent "the real nuclear renaissance."[11] Consequently, nuclear fission and new reactor designs continue to receive enormous research and development budgets in a number of countries. An increasing share of nuclear R&D funding in several countries is going towards developing and commercializing SMRs. In November 2012, the United States Department of Energy announced that as part of its SMR Licensing Technical Support Program, it would offer financial support of up to $452 million towards the development of the Babcock & Wilcox Company's mPower SMR and one more SMR design. Other countries have been following suit. The Korea Atomic Energy Research Institute is currently developing the SMART (System-integrated Modular Advanced ReacTor) and the Bhabha Atomic Research Centre in India has been developing an Advanced Heavy Water Reactor (AHWR). Russia is in the process of constructing a floating nuclear plant. More recently, in December 2017, the UK government announced that up to £100m will be made available for the development of SMRs, making them "the next big thing in energy."[12]

Historically too, nuclear power has been favored with considerable government largesse. Indeed, between 1974 and 2010, countries belonging to the Organization of Economic Cooperation and Development have spent 52 percent of their $482 billion in public research and development funds on nuclear power, greater than any other source.[13] This amount is more than five *times* the amount spent on renewable sources of energy such as ethanol, solar panels, and wind turbines combined.

Second, my exploration of the utopian fantasies surrounding SMRs helps give meaning to current debates over energy security, technology, and policy. Today's investment decisions in technologies like SMRs will "lock-in" future energy trajectories for decades to come.[14] Closely examining SMRs can shed light on the dramatic difficulty of this task, and help inform future strategies for managing energy transitions.

Third, and more broadly, the chapter traces the influence that fantasies have on the practice of science, in this case the scientific design of reactors and their related components. For the moment, SMRs are almost entirely a rhetorical construction. Engineers have not yet built any commercial SMRs, and the debate about them hinges entirely on visions and expectations. As Cynthia Selin put it, "the future is an active arena, one both pregnant and populated with agendas, interests, and contestations … Like the legitimization processes that occur during fact building, expectations serve a very real, very palatable role" in technological development and innovation.[15] The rhetorical experiences and fantasies associated with SMRs therefore have relevance for all those concerned about technology choice, innovation, commercialization, and the use of fantasies in influencing policy decisions.

Conceptualizing small modular reactors (SMRs)

The next part of the chapter investigates the many ways in which utopian fantasies and rhetorical visions are used to describe and promote SMRs. Before

I explore those, however, it is important to clarify what I mean by SMRs. The acronym SMR stands for two related terms. In the United States and in some other parts of the world, it stands for "small, modular reactors." While the term small may seem self-explanatory, the second term, modular, means that these reactors are to be assembled from factory fabricated modules, with each module representing a portion of a finished plant, rather than constructed on site, as most current nuclear power stations are. Modularity is also used to indicate the idea that rather than constructing one large reactor, the equivalent power output will be generated using multiple smaller reactors that allow for greater tailoring of generation capacity to demand. The International Atomic Energy Agency (IAEA), on the other hand, uses the acronym to mean "small and medium-sized reactor." For the IAEA, a "small" reactor is one having electrical output less than 300 megawatt-electrical (MWe) and a "medium" reactor is one having a power output between 300 and 700MWe. Thus, for the IAEA, any reactor with electrical output less than 700MWe is an SMR, even if not necessarily modular. In this paper, I use SMR in both senses. SMRs have electrical power outputs substantially smaller than currently operating reactors or those under construction (see Table 2.1).

While the term SMR is widely used, it actually does not represent any one kind of reactor. Rather, there are multifarious SMR designs with distinct characteristics being developed, with 39 already registered with the International Atomic Energy Agency across many different suppliers and countries, as Table 2.2 reveals. These designs vary by power output, physical size, fuel type, enrichment level, refueling frequency, site location, and spent fuel characteristics. They are also in different levels of development, with some in the process of being constructed (e.g. the Russian KLT-40 floating power plant and the Chinese HTR-PM reactor), and others that still face major technical challenges unlikely to be overcome during the next decade. This multifariousness is useful to SMR advocates by allowing them to simultaneously claim multiple desirable characteristics for SMRs in general; these characteristics would not all be realizable, even theoretically, in a single design.

TABLE 2.1 Electrical power outputs of reactors (in MWe)

	Operating	Under Construction
Pressurized water reactors	924	972
Boiling water reactors	913	1,312
Pressurized heavy water reactors	496	538
Small modular reactors	–	50–300*

Source: Modified from IAEA 2011.[16] MWe=Megawatt electric. *most designs, with some exceptions

TABLE 2.2 Common small modular reactor designs under research and development

Design	Name	Reactor type	Designer	Country	Capacity(MW(e))
Water cooled	CAREM-25	Integral pressurized water reactor	CNEA	Argentina	27
	ACP-100	Integral pressurized water reactor	CNNC (NPIC/CNPE)	China	100
	Flexblue	Subsea pressurized water reactor	DCNS, France	France	160
	AHWR300-LEU	Pressure tube type heavy water moderated reactor	BARC, India	India	304
	IRIS	Integral pressurized water reactor	IRIS	International Consortium	335
	DMS	Boiling water reactor	Hitachi-GE Nuclear Energy	Japan	300
	IMR	Integral modular water reactor	Mitsubishi Heavy Industries	Japan	350
	SMART	Integral pressurized water reactor	KAERI	South Korea	100
	KLT-40S	Pressurized water reactor	OKBM Afrikantov	Russia	35 × 2 modules barge mounted
	VBER-300	Integral pressurized water reactor	OKBM Afrikantov	Russia	325
	ABV-6M	Pressurized water reactor	OKBM Afrikantov	Russia	6 × 2 modules, barge mounted, land based
	RITM-200	Integral pressurized water reactor	OKBM Afrikantov	Russia	50

	Reactor type	Developer	Country	Power
VVER-300	Water-cooled water-derated power reactor	OKB Gidropress	Russia	300
VK-300	Simplified boiling water reactor	RDIPE Development Engineering	Russia	250
UNITHERM	Pressurized water reactor	RDIPE Development Engineering	Russia	6.6
RUTA-70	Pressurized water reactor	RDIPE Development Engineering	Russia	70
SHELF	Pressurized water reactor	RDIPE Development Engineering	Russia	6
ELENA	Pressurized water reactor	Research Russian Centre "Kurchatov Institute"	Russia	0.068
mPower	Integral pressurized water reactor	Babcock and Wilcox Generation	United States	180 × 2 modules
NuScale	Integral pressurized water reactor	NuScale Power LLC	United States	45 × 12 modules
Westinghouse SMR	Integral pressurized water reactor	Westinghouse Electric Company LLC	United States	>225
SMR-160	Pressurized water reactor	Holtec International	United States	160
HTR-PM	Pebble Bed HTGR	Tsinghua University	China	211
High Temperate Gas Cooled GT-HTR300	Prismatic Block HTGR	Japan Atomic Energy Agency	Japan	100–300
GT-MHR	Prismatic Block HTGR	OKBM Afrikantov	Russia	285
MHR-T reactor/Hydrogen production complex	Prismatic Block HTGR	OKBM Afrikantov	Russia	4 x 205.5
MHR-100	Prismatic Block HTGR	OKBM Afrikantov	Russia	25–87

(Continued)

TABLE 2.2 (Cont.)

Design	Name	Reactor type	Designer	Country	Capacity(MW(e))
	PBMR-400	Pebble Bed HTGR	Pebble Bed Modular Reactor SOC Ltd,	South Africa	165
	HTMR-100	Pebble Bed HTGR	Steenkampskraal Thorium Limited (STL)	South Africa	35 per module (140 for 4 module plant)
	SC-HTGR	Prismatic Block HTGR	AREVA	France	272
	Xe-100	Pebble Bed HTGR	X-energy	United States	35
Liquid metal-cooled fast spectrum	CEFR	Sodium-cooled fast reactor	China Nuclear Energy Industry Corporation	China	20
	PFBR-500	Sodium-cooled fast breeder reactor	Indira Gandhi Centre for Atomic Research	India	500
	4S	Sodium-cooled fast reactor	Toshiba Corporation	Japan	10
	BREST-OD-300	Lead-cooled fast reactor	RDIPE	Russia	300
	SVBR-100	Lead Bismuth cooled fast reactor	AKME Engineering	Russia	101
	PRISM	Sodium-cooled fast breeder reactor	GE Nuclear Energy	United States	311
	EM2	High temperature helium-cooled fast reactor	General Atomics	United States	240
	G4M	Lead-bismuth cooled fast reactor	Gen4 Energy Inc.	United States	25

Source: Author's compilation, modified from International Atomic Energy Agency 2014[17]

Utopianism and fantasy in energy systems

As a guiding lens for my analysis of SMRs, I rely on concepts arising from an examination of utopianism and technology. At its most basic level, a utopia here is "technological pursuit of salvation."[18] Berkout adds that "utopias represent examples of radical and more fully worked visions of the future. Their aim is to break the bonds of the existing order, to exemplify an alternative order and to inform collective action in pursuit of that order."[19]

For Segal, a "technological utopia" is a mode of thought that vaunts technology as bringing about some sort of perfect society.[20,21] His historical examination of technological utopianism suggests that it reflects radicalism, comprehensiveness, and seriousness. Technological utopias are *radical* improvements of conditions compared to present. They are *comprehensive*, seeking to change most and often all areas of society. And they are *serious*, often seen as a response to some pressing real-world problem, a finding confirmed by Halley and Vatter.[22] Utopias thus tend to be apolitical as well, presented by advocates as tools for societal enhancement rather than personal gain. Segal traced 25 historical accounts of technological utopias written between 1883 and 1933 in the United States and found that all of the visions were remarkably similar: all relied on technological advances, all saw visions through lenses of their own time, all followed a series of common structures or tropes:[23]

- Evolutionary utopias depicted a future world that must be planned for and gradually shifted towards, like Darwinian evolution they would not occur overnight, and required work;
- Efficient utopias emphasized a technology's ability to reap the greatest savings, accomplish the greatest good for society, and/or minimize waste;
- Rational utopias would enable humans to (finally) conquer their emotions, especially escaping fear, despair, or loneliness;
- Anthropocentric utopias enabled humans to further conquer nature, control the environment, and dominate the physical world;
- Individualistic utopias emphasized the enhancement of household or individual traits such as wealth or education, with self-improvement taken as a proxy for improving the world (sometimes even at the expense of the world);
- Imperial utopias would involve displacing another culture or imposing a utopia on a class of people or place in order to make them healthier, happier, or richer (by force).

Segal also noted a sad undertone to the architects of these utopian visions: many individuals were driven to despair (and some committed suicide) after they were unable to introduce the reforms necessary to bring about their plans.

Such utopian visions involve not only experts tasked with constructing a particular technological world but also the imagination of appropriately behaved publics that are expected to live in it.[24,25] Moreover, utopianism has played an integral role in elevating intellectual elites so that they can accomplish particular "transcendental visions" and then reorder entire societies through revolution— transformation that often has material as well as spiritual dimensions.[26] The utopian elements of fantasies have even led proponents and sponsors to exaggerate potential benefits and downplay risks of many different technologies.[27,28]

One certainly sees utopian undertones in both previous visions of electricity as well as nuclear power specifically. At the turn of the twentieth century, for example, many commentators believed that electricity would facilitate a future age of utopian dimensions. As Marvin masterfully demonstrates, these utopias involved:[29]

- The elimination of crime: electric lighting would make streets safer to walk; electric devices would be used to foil or frighten criminals, gather evidence, and alert authorities to crimes in progress; electric chairs (and the death penalty) would also serve as a deterrent to crime;
- A transformation of work: electrical motors would greatly ease the strains of manufacturing work and eliminate drudgery, leading to more efficient economic systems, with greater versatility and efficiency; one extreme variant predicted a future of robots, electrical machines that could closely resemble people, becoming the most obedient servants;
- An extension of human lifespan: electricity would change medicine and healthcare given that it was believed to be a healing agent, with special electrical and vital powers;
- The maintenance of military superiority: electric telephone communication, high potential dynamos and electric engines would become capable of delivering electric shocks to approaching enemies; mechanical and automated weapons that could kill without compromise or compassion were also connected to electricity;
- The enhancement of entertainment: electricity was prized for its ability to be utilized in practical jokes as well as spectacles, it could offer illumination to fairs and light shows, dazzle audiences, and increase the effectiveness of marketing and advertising;
- The preservation of democracy and peace: electricity was thought by some to render politics obsolete, as it would result in globalized communication patterns and an electrical world void of cultural differences;
- The provision of pleasure: a final utopian theme was that an electrical world offered pleasure and "inexhaustible novelty;" fulfilling every need since all pleasures could be met by pushing a button, a theme partially explained the allure of electrical jewelry and also "electric girls" who adorned themselves with lightbulbs.

Marvin warns that each of these utopias tended to rely on a "cognitive imperialism," a meta-vision of a world efficiently administered by Anglo Saxon technology with Anglo Saxon people.[30] They all depicted Western civilization at the center of a stage play for which the rest of the world was an awestruck and passive audience.

To be fair, these utopian fantasies of an electrical world were not uniform, nor did they remain uncontested.[31] Given that the properties of electricity were not that well understood at the time (and electricity networks only at nascent stages of development), others did take discomfort with the menace and risk of new electrical technologies. Many believed that use of electricity contributed to "charging" the world, or accumulating electricity so that the earth could explode, or that increasing use of electricity was connected to greater amounts of lightning in the atmosphere. Some even argued that the use of electricity foretold the apocalypse, and that society would end within the next few decades. While neither the utopian or dystopian visions of electricity were fully reached, many of their themes laid a foundation for a consumer age. They supported a general idea of rapid progress, a time of constant electrical invention, an era of dazzling and spectacular new technologies. This conditioned views of the future, and created expectations about what was possible and achievable, breaking previous limits and standards.

This expansion of utopianism becomes clear when we consider a slew of nuclear fantasies and visions that begun to emerge a few decades after the large-scale introduction of electricity to society. The same month that the atomic bombs were dropped on Hiroshima and Nagasaki in 1945, the pocketbook *The Atomic Age* was published and widely read. The book depicted a future world where coal and petroleum would go unused, existing hydroelectric facilities would be rendered as "obsolete as the stagecoach" was. To give the general public some feeling for the vast amounts of energy soon to be theirs, the authors of *The Atomic Age* heralded the untapped atomic power within everyday objects: a jug of lemonade had enough energy to heat 100 million tons of water; a handful of snow could power an entire city; the energy in a small paper railway ticket was sufficient to power a heavy passenger train several times around the earth.[32]

Four statements nicely capture the utopian optimism over atomic energy during this period. David Dietz, a science writer, wrote in 1945 that:

> Instead of filling the gasoline tank of your automobile two or three times a week, you will travel for a year on a pellet of atomic energy the size of a vitamin pill. The day is gone when nations will fight for oil. The world will go permanently off the gold standard once the era of Atomic Energy is in full swing. With the aid of atomic energy, the scientists will be able to build a factory to manufacture gold. No baseball game will be called off on account of rain in the Era of Atomic Energy. No airplane will bypass an airport because of fog. No city will experience a winter traffic jam because of snow. Summer resorts will be able to guarantee the weather and artificial suns will make it as easy to grow corn and potatoes indoors as on a farm.[33]

Glen Seaborg, the scientist who co-discovered plutonium and also chaired the United States Atomic Energy Commission (AEC), noted that:

> The future of civilization is in the hands of the nuclear scientists who form the elite team that will build a new world through technology ... There will be nuclear powered earth-to-moon shuttles, nuclear-powered artificial hearts, plutonium-heated swimming pools for SCUBA divers, and much more. My only fear is that I may be underestimating the possibilities.[34]

Robert M. Hutchins, the president of the University of Chicago, stated in 1946 that nuclear power would make:

> Heat so plentiful that it will even be used to melt snow as it falls. A very few individuals working a few hours a day at very easy tasks in the central atomic power plant will provide all the heat, light, and power required by the community and these utilities will be so cheap that their cost can hardly be reckoned.[35]

Captivated by such optimism, Lewis Strauss, the chairperson of the AEC in 1954, remarked that atomic power would usher in an age where:

> It is not too much to expect that our children will enjoy in their homes electrical energy too cheap to meter, will know of great periodic regional famines in the world only as matters of history, will travel effortlessly over the seas and under them and through the air with a minimum of danger and at great speeds, and will experience a lifespan far longer than ours as disease yields and man comes to understand what causes him to age.[36]

By the mid-1950s and 1960s the utopian spirit associated with nuclear power promotions was omnipresent. Posters heralded the dawn of a new age. Even Walt Disney was not immune to the technological promise of nuclear power, producing a book and animated television show entitled "Our Friend, the Atom." Isaac Asimov predicted in 1964 that:

> The appliances of 2014 will have no electric cords, of course, for they will be powered by long- lived batteries running on radioisotopes. The isotopes will not be expensive for they will be by- products of the fission-power plants which, by 2014, will be supplying well over half the power needs of humanity.[37]

Not to be outdone, in 1967 Westinghouse designed a series of pamphlets on nuclear power which declared "we have found what might be called perpetual youth."[38]

These visions, simply put, were pervasive and captivating. As David Lilienthal, chair of the AEC from 1947 to 1949, put it years later:

> The basic cause [of the inflated hopes for atomic energy] was a conviction, and one that I shared fully and tried to inculcate in others, that somehow or the other the discovery that had produced so terrible a weapon had to have an important peaceful use. Such a sentiment is far from ignoble … Everyone … wanted to establish that there is a beneficial use of this great discovery. We were grimly determined to prove that this discovery was not just a weapon. This led, perhaps, to wishful thinking.[39]

The remainder of this chapter shows that such fantastic utopias (and wishful thinking) are not a relic of the past. They continue to exist about state-of-the-art SMR designs within the contemporary physics community.

Research design and empirical strategy

Drawing from previous work,[40] M. V. Ramana and I adopted a specific schema to analyze the ways in which SMRs are conceived of and described. Given that the goal was to identify the prevalence of fantasies among practicing scientists and engineers (mostly physicists), rather than popular writers or politicians, we focused the literature review entirely on academic and technical articles. This approach is similar to methodologies employed by McDowall and Eames[41] and Sovacool and Brossmann[42] to illustrate visions articulated by advocates of the hydrogen economy. The sample of the literature was limited to English publications only.

The investigation began by recording visions of SMRs as articulated by their proponents in two sets of literature: peer-reviewed energy and nuclear power journals, and International Atomic Energy Agency (IAEA) publications. Peer-reviewed academic studies were identified by searching the ScienceDirect database for the phrase "small modular reactor" in a study's title, abstract, or keywords published from January 2002 to September 2012. ScienceDirect was chosen because it is home to most of the scientific journals dealing with nuclear power, including the *Annals of Nuclear Energy, Annals of Nuclear Science and Engineering, International Journal of Radiation Applications and Instrumentation, Journal of Nuclear Energy, Journal of Nuclear Materials, Nuclear Physics, Progress in Nuclear Energy*, and *Progress in Particle and Nuclear Physics*, among others. The initial search identified 67 articles, of which 42 were actually about SMRs for the power sector and 34 articulated some type of rhetorical vision.

To supplement these academic papers, 38 reports published from 2004 to 2012 were collected from the IAEA website. This choice was based on the fact that the IAEA, with more than 100 member countries and a staff of 2,300 professional personnel, represents "the world's central intergovernmental forum for scientific and technical co-operation in the nuclear field."[43] The inclusion of IAEA documents gives my pool of studies a more "global" character representative of the multitude of countries currently pursuing SMR research.[44] Of these 38 reports, 26 articulated some type of rhetorical vision in association with SMRs. Figure 2.1 depicts my sampling process.

Appendix 2.1 illustrates some of the features of the final sample of 60 articles. Academic articles tended to come from the technical literature on nuclear energy, notably journals such as *Nuclear Engineering and Design* and *Progress in Nuclear Energy*. These articles were written predominately by authors affiliated with universities, energy companies, consulting firms, and government sponsored research laboratories. While their institutional affiliations represented 10 countries, almost one-third (29 percent) had primary authors from the United States. Interestingly, however, multiple studies were also from institutions in France, Japan, Russia, and South Korea, implying that visions of SMRs, and not just technology development programs, are widespread. Publications from the IAEA also featured authors from several countries.

Admittedly, there is something of a disconnect in the historical theme of utopianism, most studied (and seemingly widespread) in popular culture, and thus popular mass media sources, and the exclusive focus in this chapter on scientific papers written for scientists and published in English. But the idea is to show how fantasies can proliferate even into technical journals and scientific thinking, and to reveal how fantasies circulate beyond the realm of science fiction or popular discourse alone.

Five utopian SMR visions

The most recent, and current, wave of interest in SMRs dates back to the early 2000s. The problem, as laid out by analysts from the IAEA's Department of

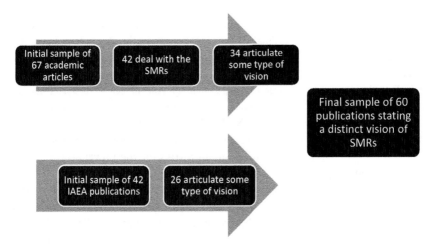

FIGURE 2.1   Literature selection process for academic and IAEA SMR studies
Source: Author, modified from Sovacool, B. K., and M. V. Ramana 2015, January[45]

Nuclear Energy, was that "quite simply, over the last 15 years, nuclear power has been losing market share badly in a growing world electricity capacity market."[46] Their diagnosis:

> the main reason for this stalemate is that we, in all our doings, continue to rely on nuclear technology developed in the 1950s, which had its roots in military applications which cannot exclude absolutely the possibility of a severe accident and which has reached its limits from an economic point of view.[47]

As the way forward, the IAEA suggested developing innovative new reactor designs, chiefly of the SMR variety. Since then, the discourse about SMRs has moved from SMR characteristics being "desirable" to the claim that they will definitely be "achieved."

The idea of developing SMRs has secured support from a multitude of individuals and organizations for a variety of reasons. Indeed, my sample of SMR literature depicted a diverse number of rhetorical visions, ranging from cost competiveness with existing sources of energy, the jobs and technological learning that would accompany mass production of SMR units, the ability to undertake "advanced oil recovery" through unconventional reserves such as oil shale and tar sands, large-scale hydrogen production, and the creation of process heating for chemical and manufacturing processes, among others.

In this section I don't touch on all these visions for want of space, and also because they are arguably less "utopian," "fantastic" and "exciting" than five that stood apart from the rest. I believe that the allure of SMRs arises in part from their purported ability to deliver amazing advances in how we produce and consume energy, in how we address some global problems, even in how we determine our next phase of outer space exploration, with five striking visions summarized by Table 2.3, which shows the specific elements of each rhetorical vision, and Table 2.4, which shows their frequency. The most popular vision concerned risk-free energy (presented in 43 studies) followed by self-energization (24), environmental nirvana (22), water security (21), and space exploration (2).

While functionality is apparent in essentially all the papers analyzed, utopianism is more an underlying ideal behind each vision. In this ideal world, SMRs would generate plentiful energy of multiple kinds (electricity, heat), providing the necessary means for a life of comfort for all people by meeting various needs (lighting, temperature control, drinking water, scarce minerals) and without any environmental externalities or cause for concern about accidents.

Risk-free energy

By far the most prevalent rhetorical vision concerning SMRs, present in more than 60 percent of my sampled studies, relates to their ability to generate energy without risk, with extreme reliability and perfect safety. Characters for this vision

TABLE 2.3 Five rhetorical visions associated with SMRs

Rhetorical vision	Explanation	Storyline or characters	Symbolic cue(s)	Audience
Risk-free energy (n=43)	SMRs can produce energy with perfect reliability and complete safety	Improperly trained and error-prone human operators as well as terrorists and potential saboteurs seeking to cause nuclear catastrophes	"passive safety," "inherently safe," "sure protection," "safety by design" "operator free"	Investors, financiers, suppliers, operators, concerned citizens (often living near existing or proposed nuclear facilities)
Indigenous self-energization (n=24)	SMRs can provide energy autonomy and self-determination for remote areas and developing countries	Those without access to modern energy services, rapid population growth	"Remote village communities," "limited grids," "unsophisticated grids," "just-in-time capacity growth"	Philanthropists and potential investors
Environmental nirvana (n=22)	SMRs can deliver clean and plentiful electricity in a carbon constrained future	The impending specter of climate change and environmental degradation as well as the difficulty of storing nuclear waste	"Waste-free energy," "carbon free energy," "zero carbon energy source"	Environmentalists and concerned citizens (often living near existing or proposed nuclear facilities)
Water security (n=21)	SMRs can desalinate water needed to avert global water crises	People living in water stressed and water scarce areas	"Universal access to water," "non-electric markets," "process heat applications"	Politicians and environmentalists
Space exploration (n=2)	SMRs are needed for the exploration of space and interstellar propulsion	The universal call for scientific experimentation and discovery	"Lunar outputs," "Mars mission"	Scientists, space enthusiasts

Source: Author, modified from Sovacool, B. K., and M. V. Ramana 2015, January[48]

TABLE 2.4 Frequency of SMR rhetorical visions

Study	Risk-free energy	Self-energization	Water security	Environmental nirvana	Space exploration
Bortot et al. (2011)		✓			
Carelli et al. (2004)	✓				
Carelli et al. (2010)				✓	
Choi et al. (2011)	✓	✓		✓	
Chung et al. (2004)	✓		✓		
Ding and Kloosterman (2011)	✓	✓			
El-Genk et al. (2004)	✓	✓	✓		
Filho (2011)	✓				
Fuls and Mathews (2007)	✓				
Hatton et al. (2009)					✓
Hibi et al. (2004)	✓				
Hong et al. (2008)				✓	
Ingersoll (2009)	✓	✓	✓	✓	
Kang et al. (2012)	✓				
Karahan (2010)	✓				
Kessides (2012)	✓	✓	✓	✓	
Lee et al. (2012)	✓	✓	✓		
Lim et al. (2011)		✓			
Matsumura et al. (2005)	✓			✓	
Sahin and Sefidvash (2008)	✓			✓	
Sambuu et al. (2012)	✓	✓		✓	
Shropshire (2011)		✓		✓	
Slessarev (2007)	✓				
Slessarev (2008)				✓	
Tominaga et al. (2005)	✓				
Toshinsky et al. (2011)	✓			✓	
Vujic et al. (2012)		✓		✓	
Wallace et al. (2006)	✓	✓			
Yang et al. (2006)	✓		✓	✓	
Yoon et al. (2012)		✓			
Zhang et al. (2007)	✓				
Zheng et al. (2012)	✓				
Zrodnikov et al.(2008)	✓	✓		✓	
Zrodnikov et al. (2011)	✓	✓	✓		
International Atomic Energy Agency (2004)	✓	✓	✓		
International Atomic Energy Agency (2005)	✓	✓	✓	✓	
International Atomic Energy Agency (2006a)	✓	✓	✓	✓	
International Atomic Energy Agency (2006b)	✓				

(*Continued*)

TABLE 2.4 (Cont.)

Study	Risk-free energy	Self-energization	Water security	Environmental nirvana	Space exploration
International Atomic Energy Agency (2007a)	✓	✓	✓		
International Atomic Energy Agency (2007b)			✓		
International Atomic Energy Agency (2007c)	✓	✓	✓	✓	✓
International Atomic Energy Agency (2009a)			✓	✓	
International Atomic Energy Agency (2009b)		✓			
International Atomic Energy Agency (2009c)	✓				
International Atomic Energy Agency (2009d)	✓		✓		
International Atomic Energy Agency (2009e)	✓			✓	
International Atomic Energy Agency (2009f)			✓		
International Atomic Energy Agency (2010a)			✓		
International Atomic Energy Agency (2010b)	✓			✓	
International Atomic Energy Agency (2010c)	✓			✓	
International Atomic Energy Agency (2011)			✓		
International Atomic Energy Agency (2012a)	✓			✓	
International Atomic Energy Agency (2012b)			✓		
International Atomic Energy Agency (2012c)	✓				
International Atomic Energy Agency (2012d)	✓		✓		
Kuznetsov (2006)	✓	✓	✓		
Kuznetsov (2007)	✓				
Kuznetsov (2008)	✓	✓			
Kuznetsov (2009)	✓				
Kuznetsov et al. (2009)		✓			

Source: Author. Note: See "Appendix 2.1" to match the IAEA citations in this table with the full titles of those reports

are both improperly trained human operators in addition to would-be terrorists and saboteurs. The most prominent symbolic cue of SMRs being "inherently safe" was mentioned in more than a dozen studies within my sample. Investors, financiers, suppliers, operators, and ordinary citizens comprise the likely audience for this vision. Given the heightened concerns about nuclear accidents following Fukushima, the functionality of this vision is obvious.

The risk-free energy vision begins by acknowledging that existing nuclear power plants are prone to accidents from a variety of causes such as Three Mile Island (operator error), Chernobyl (a mishandled safety test), and Fukushima (an earthquake and tsunami). The IAEA recognizes that natural events beyond those that contributed to Fukushima, such as extreme metrological conditions including changes in temperature or subsurface freezing, floods from a variety of sources, cyclones, lightning, landslides, and avalanches, can contribute to the risk of accidents at operating reactors. Accidents can be human induced from aircraft crashes, explosions, bombings, electromagnetic interference, and other forms of sabotage. They can be caused by "internal events," such as rupturing pipelines, blocked valves, and malfunctioning equipment, as well as "external events," such as flooding, earthquakes, and tsunamis.[49]

Proponents of the risk-free SMR vision comment that ensuring adequate safety at these existing plants is impossible, with reactor safety and control imperfect due to the "confined capability of the first nuclear power plant generations to withstand severe accidents."[50] Authorities, in other words, have been "obliged" to "compromise" safety standards in the face of inferior technology. SMRs, by contrast, can benefit from a "fresh safety strategy"—in essence, the functional element of the vision—that provides "sure protection against all severe accidents." As some of the academic studies state repeatedly, the "safety-by-design" approach facilitated by SMRs can focus on "eliminating by design the possibility for an accident to occur, rather than dealing with its consequences."[51] The end result is SMRs possess "inherent safety properties (deterministic elimination of severe accidents) that enable [them] to assure a high level of social acceptability."[52]

The stated justification for this vision lies in the combination of passive safety features, multiple defensive barriers, and designs that explicitly prioritize safety over all other criteria (such as economics or, say, carbon intensity). Because of their smaller physical size, SMRs can feature "larger reactor surface-to-volume" ratios that can facilitate passive removal of decay heat.[53] SMRs can operate at lower power densities, have lower fuel inventory, and can be placed underground, all reducing the chances of an accident and containing its impacts should it occur.[54]

This risk-free energy vision has broad support across cultures, institutions, and reactor types, implying that it is both compelling and pervasive (affirming, also, the utopian dimension of the vision). Chinese researchers argue that design features in SMRs allow them to "solve" the accident problem and "make sure" that "reactors will not melt."[55] A Russian research team speaks of "inherent self-protection" and "passive safety," pointing to calculations that show that "the maximum admissible temperature of fuel elements' claddings is not exceeded" for a "variety of postulated accidents and abnormal situations," with the proviso that "no other potentially realized scenarios of accidents which can result in hazardous consequences have been found."[56] Though they were writing before the Fukushima accident, Japanese researchers assure us that their SMR design's chosen "configuration eliminates a pressurizer, reactor coolant pumps and all loop pipes, making the plant system drastically simple and eliminating accidents which cause fuel failure."[57] A South African research team writes about "inherently safe design" which "renders obsolete the need for safety back-up systems and most aspects of the off-site emergency plans required for conventional nuclear reactors," promising that their SMR design can "withstand" aircraft impact and tornados and would be "resilient" to explosions from terrorists and sabotage.[58] Korean scientists intone that in their design, "the possibility of a large break loss of coolant accident is eliminated."[59] An American scientist writes that SMR designs have "eliminated accident vulnerabilities,"[60] while a separate American research team argues that SMRs "could make extensive use of passive cooling techniques during normal operation and for the removal of decay heat after shutdown, precluding the likelihood of severe core damage accidents and/or radioactivity release."[61] The IAEA points out that most "innovative reactors aim to eliminate the need for intervention in public domain (outside the plant boundary) through the use of enhanced passive safety features in their design."[62] Assuming that this aim is achieved, there would be no need to require emergency planning procedures.

Indigenous self-energization

The second most popular rhetorical vision among my sample relates to indigenous self-energization, and it espouses SMRs as a way to empower communities and emerging economies with energy autonomy and self-determination. This vision, its functionality and utopian dimensions, takes many forms. The most popular states the acceptability of SMRs for rapidly developing economies, such as those in Brazil, China, India and South Africa. Another form sees SMRs as vital for the needs of developing countries with small electrical grids and unsophisticated infrastructure, such as East Timor or Zanzibar. One sees SMRs as key to electrifying small off-grid villages, towns, and sparsely populated places such as the Aleutian islands in Alaska or Papua New Guinea. One envisions SMRs as supplying energy to off-grid mining villages requiring intensive energy resources like those throughout Mongolia and Australia.[63] One variant argues that SMRs are ideal for countries with large territories but relatively small and dispersed populations.[64]

Characters for this vision, independent of its variants, are those without access to modern energy services afflicted by energy poverty and rapid population growth. Symbolic cues include phrases such as "remote village communities" and "just-in-time capacity." The primary audience for this vision appears to be people and agencies interested in global economic development.

The best articulation of this vision comes from the IAEA itself. The organization states in their flagship *Nuclear Technology Review* that:

> Growing populations, plus increasing urbanization and growing per capita energy use driven by development, may create a market for SMRs because of limited grids in many countries and limited investment capabilities. By 2015, more than 370 cities in Asia, Africa, and Latin America are expected to have more than one million people each; collectively, these cities would account for 1.5–2 billion people. To accommodate rapid demand growth where initial grids and financing are limited, a "just-in-time" capacity growth plan might be appropriate, with incremental capacity additions as the population grows, as per capita energy use increases, and as a city becomes wealthier. SMRs could meet the needs of these emerging energy markets where the industrial and technical infrastructure is generally poor, if they are designed to be easily expandable into clusters comprising ever-larger power installations.[65]

Other IAEA documents inform us that SMRs are especially attractive for "developing countries with small electric grids, insufficient infrastructure, and limited investment capability" and "today's developing countries, which often have insufficient infrastructure and small electricity grids"[66,67] Not to be outdone, Vladimir Kuznetsov from the IAEA's Nuclear Power Technology Development Section tells us that:

> The role of SMRs in global nuclear energy system could then be to increase the availability of clean energy in a variety of usable forms for all regions of the world, to broaden the access to clean and affordable and diverse energy products and, in this way, to contribute to the eradication of poverty and, subsequently, to peace and stability in the world.[68]

The IAEA states that Russian nuclear designers have developed three reactors that are to be "barge-mounted, complete power plants which can be towed from the factory to a water accessible site, moored in a pre-prepared lagoon, and connected to a localized grid."[69] These designs, we are assured, "could support electrical needs for off-grid towns of up to several hundred thousand populations," and "are also properly sized for support of industrial operations at remote, water-accessible locations."

The IAEA is not alone in subscribing to a vision of self-energization. A research team at the University of California Berkeley writes that "small reactors are ideal for providing the electricity to countries with small, limited, or distributed electricity grid system as well as for countries with limited financial resources for investment in large nuclear power plants."[70] Researchers at the University of New Mexico write that:

> The populations in underdeveloped countries and in small remote communities, with small or nonexistent electrical grids, inadequate socioeconomical infrastructure, limited or diminishing water resources and access to fossil fuel, represent more than 80 percent of the world's total population. Future energy needs for this population of more than 4.5 billion could effectively be met using very small, modular and cogeneration nuclear reactor power plants, each generating a few to tens of megawatts of electricity and heat for a variety of applications.[71]

A similar monograph from a researcher at Oak Ridge National Laboratory comments that:

> the enhanced safety, security, and operational features of [SMRs] make them well suited to support a global expansion of nuclear energy and may, in fact, be the enabling factor for smaller, developing countries. It is conceivable that the first unit of the new breed of deliberately small reactors will be built first in a smaller country that has no other options.[72]

Scientists at the State Scientific Center of Russian Federation Institute for Physics and Power Engineering argue that SMRs provide options that allow them to be deployed in "developing countries that have no deployed networks for electrical energy transmission and distribution."[73] A team from Japan's Hitachi Power Systems informs us that "the needs of medium and small sized reactors have become greater in foreign countries where electric grid systems are weak" and their company has developed an SMR that can "be introduced in the market in near future."[74]

Some engineers and manufacturers are so confident in the truth behind this vision that they are building SMRs explicitly for developing countries without well-developed electricity infrastructures. One prototype, cleverly called "SUPERSTAR" (for SUstainable Proliferation-resistance Enhanced Refined Secure Transportable Autonomous Reactor), is "intended for international or remote deployment," and sized with smaller power levels to match the "smaller demand of towns or sites that are either off-grid or on immature local grids, being right-sized for growing economies and infrastructures of developing nations."[75]

One special characteristic that many reactor designers seek is "operator free" performance, which intertwines the vision of risk-free energy with that of self-energization. Scientists and engineers from the Kurchatov Institute point out that this "feature becomes very important if such plants are to be used for heat and electricity supply to small towns in the remote areas with no centralized power supply."[76] In other words, the reactor is envisioned as being appropriate to remote areas because it will not need trained operators to move to these locations. To make people feel comfortable with the idea that there is a reactor among them uncontrolled by any operators, analysts from Central Research Institute of Electric Power Industry of Japan point out that they aim "to exclude human errors in the reactor operation."[77]

Environmental nirvana

The third most frequent rhetorical vision, one of environmental nirvana, depicts SMRs as a "low-carbon" or "no-carbon" energy option that can produce energy cleanly without waste. The compelling, functional, utopian narrative behind this vision are all of the calamities to be expected with impending climate change and the destruction of the environment, including rises in sea level and more frequent storms, as well as the difficulties of storing nuclear waste. Symbolic cues for this vision include terms such as "waste-free" and "zero carbon" energy. Environmentalists and concerned citizens (those likely to oppose nuclear power plants for waste concerns) are the primary audience for this vision.

This vision is particularly persuasive because it takes two of the most common reasons for opposing nuclear power—its poor environmental record and its legacy of long-lived radioactive waste—and turns them into advantages. The IAEA, for example, admits that existing reactors can damage the environment throughout their fuel cycle, from the uranium mine and enrichment facility to the power plant itself, waste facilities, and auxiliary buildings. Moreover, the generation of nuclear electricity involves multiple "stressors" to the environment, including radioactivity, chemical toxins, pathogens, heat, erosion, noise, and the depletion of resources. These "stressors" can contaminate people, communities, and biodiversity through a variety of pathways, including the air (through inhalation or contact), surface water (ingestion or contact), groundwater (ingestion or contact), and habitat (terrestrial or aquatic).[78] Similarly, studies from my sample posited that the issue of nuclear waste represented a "painful point" for the industry,[79] and the nuclear industry's "future has been clouded" by, inter alia, the "challenges of radioactive waste disposal."[80]

SMRs, by contrast, offer the ability to tackle these challenges and avert environmental destruction. The IAEA stipulates that SMRs in particular are needed to meet "large increases in energy needs in the century ahead" with "a significant impact on global goals for reduced greenhouse gas emissions for a cleaner environment."[81,82] As another IAEA publication goes on to state:

> There is increasing interest in SMRs, which offer potential advantages [since they] expand the low emission benefits of nuclear power beyond electricity production. Half of our nation's carbon emissions come from the transportation and industrial sectors, but nuclear power has not played a significant role in these sectors. As the transportation sector begins to use more plug-in hybrids and electric vehicles, nuclear power

can help meet the additional demand for low carbon electricity production.[83]

Studies in the academic literature echo these sentiments, and note that "[SMRs] can play a very significant long-term role for meeting the world's increasing energy demands, while simultaneously addressing challenges associated with global climate and environmental impact" and that "renewed interest in SMRs is driven by low carbon."[84,85]

In terms of nuclear waste, some SMRs offer the vision of waste-free energy. One study states, the "elimination of long-lived radioactive wastes" could be "quite realistic" with SMRs, leading to a "long-lived waste free strategy."[86] In parallel, the final report of an IAEA-coordinated research project declared that SMRs can entirely "eliminate the obligations of the user for dealing with fuel manufacture and with spent fuel and radioactive waste."[87] These claims clearly serve a major function in promoting SMRs because long-lived radioactive waste is seen by many as the Achilles heel of nuclear power, but as discussed later, this functionality comes at the cost of making it more difficult to meet other goals.

Water security

The fourth most frequent rhetorical vision sees SMRs as instrumental in alleviating global water shortages for billions of people. As the IAEA succinctly states, "the desalination of seawater using nuclear energy is a feasible option to meet the growing demand for potable water."[88] Characters for this vision are people living in water stressed or water scarce areas, and symbolic cues include phrases such as "non-electric markets" and "process heat applications" for SMRs. The primary audience for this vision is politicians and environmentalists.

The best example of this vision, again, comes from IAEA publications, which note that nuclear desalination has existed for about six decades—the facilities in Table 2.5 have a collective experience of 200 reactor years—but has not yet achieved wider application. Those adhering to this vision hope that SMRs will provide the critical momentum needed to push nuclear desalination into the mainstream so that global water scarcity issues can be addressed. As one IAEA publication notes:

> Water, energy and environment are essential inputs for the sustainable development of society ... Less than 0.08 percent of the world's water is thus readily accessible for direct human use, and even that is very unevenly distributed ... The desalination of seawater using [SMRs] is a feasible option to meet the growing demand for potable water.[89]

Another IAEA report elaborates that:

> Nuclear energy can also make a substantial contribution to a critical non-energy ambition—universal access to plentiful fresh water. Currently about

TABLE 2.5 Global nuclear desalination capacities

Facility	Location	Gross power (MWe)	Capacity (m³ per day)
Shevchenko*	Aktau, Kazakhstan	150	800,000 to 145,000
Ikata-1 and Iktat-2	Ehime, Japan	566	2,000
Ikata-3	Ehime, Japan	890	2,000
Ohi-1 and Ohi-2	Fukui, Japan	2,350	3,900
Ohi-3 and Ohi-4	Fukui, Japan	1,180	2,600
Genkai-4	Fukuoka, Japan	1,180	1,000
Genkai-3	Fukuoka, Japan	2,360	1,000
Takahama-3 and Takahama-4	Fukui, Japan	1,740	1,000
Nuclear Desalination Demonstration Plant	Kalpakkam, India	170	6,300
Low-Temperature Evaporation Plant	Trombay, India	40	30
Diablo Canyon	San Luis Obispo, United States	2,200	2,180

Source: IAEA-TECDOC-1639 2010.[92] *Shut down in 1999

2.3 billion people live in water-stressed areas and among them 1.7 billion live in water-scarce areas, where the water availability per person is less than 1000 m³/year. By 2025 the number of people suffering from water stress or scarcity could swell to 3.5 billion, with 2.4 billion expected to live in water-scarce regions … Nuclear desalination is most attractive in countries that both lack water and have the ability to use nuclear energy such as China, India and Pakistan. These three countries alone account for about 40 percent of the world's population, and thus represent a potential long-term market for nuclear desalination. The market will expand as other regions with high projected water needs, such as the Middle East and North Africa, increase their nuclear expertise and capabilities.[90]

Perhaps because of these potential benefits, the IAEA launched the International Nuclear Desalination Advisory Group in 1997 "to advise the agency in nuclear desalination related studies and actions." Even those outside of the IAEA have become enrolled in this vision, with one peer-reviewed study commenting that "the continuous increase in the world's population and decrease in fresh water resources, representing less than 3 percent of the total water inventory on Earth, dictates the necessity to develop [SMRs] for both electricity and fresh water production."[91]

Similar to the indigenous self-energization vision, a multitude of arguments are presented for how and why SMRs will become vital in the global campaign to provide access to clean water.[93] SMR desalination is lauded for its ability to stimulate development and improve the resilience of communities. As the IAEA informs us:

> [SMR] desalination in general is expected to have a positive impact on
> society by making more fresh water available and opening up opportunities
> for more development related activities ... Enhanced economic activity
> after provided water availability due to desalination may cause immigration
> to the community which enjoys it, and as a consequence stimulate residen-
> tial and industrial development.[94]

Access to SMR-based desalination, furthermore, can "transform" cities and industries with "profound" effects; one example if the city of Aktau, Kazakhstan, where
"nuclear desalination was used to support the growing city and industry in the area
rich with minerals" and "transformed" a "desert-like area" into a "developed coast,
further stimulating economic as well as population growth."[95] Desalination is
praised in Spain for enabling the southern region of Almeria to bolster food
production and commodity exports, and for achieving "food independence" in
Saudi Arabia through increased production of wheat.[96]

   As with the indigenous self-energization vision, manufacturers have endorsed
this vision as well, designing special purpose SMR facilities for desalination. Such
designs modify "ordinary" SMRs so that reactors remain sealed for 15 to 20 years
using mixed oxide or nitride fuels, being refueled with another complete core so that
operation can be autonomous, load-following, and continuous.[97] South Korea, for
instance, is working on a 330 MW(th) reactor "intended for seawater desalination,"
whose design has been recently certified.[98] Its developers market it as a small nuclear
reactor for "diverse utilization" that includes not just "seawater desalination" but also
"power generation, district heating, and ship propulsion."[99] In Indonesia, the IAEA
has sponsored a "technical cooperation project" to examine the "economic viability
of construction of a nuclear desalination plant ... to support industrialization of the
Madura Region," a facility expected to come online in 2018 when it is hoped to
supply 40,000 tons of potable water per day.[100] Russian scientists are working on
a version of their VBER-300 SMR to be used as a "small-to-medium power source"
for "floating nuclear power plants or desalination complexes."[101]

Space exploration

The final SMR vision within my sample, albeit rare,[102] is perhaps the most
fantastic. Expressed by two scientists working at the Institute for Space and
Nuclear Power Studies, and Chemical and Nuclear Engineering Department at
the University of New Mexico, this vision argues that SMRs, in this particular
case "fast spectrum space reactors," are needed to operate interstellar ships
traveling with robots to Mars and beyond.[103] Another paper within my sample
also informs us about the RAPID (Refueling by All Pins, Integral Design) sodium
cooled reactor that generates 10,000 kW(th) (1,000 kW(e)), an "operator-free fast
reactor concept designed for a lunar based power system."[104]

   The primary narrative and storyline of this vision relates to the human need to
explore and discover the universe and to conduct scientific experiments, with

terms such as "lunar outpost" and "Mars mission" serving as symbolic cues. The primary audience for this vision is likely other scientists and space enthusiasts. But the vision also attempts to appeal to a larger set of stakeholders by motivating the erection of lunar outposts as necessary for "industrial activities, such as the recovery of minerals, indigenous resources and the production of propellant for subsequent travel to Mars;" essentially tying SMR development to the colonization of other planets.[105]

This vision, while it is clearly the least prominent of the five, affirms the value of "compact and lightweight nuclear reactor power systems ... supplemented by photovoltaic arrays" for the energy needs of ships, space stations, and extraterrestrial outposts on places such as the Moon and Mars. Without ignoring the widely prevalent and established use of solar photovoltaics in outer space, one study nevertheless tries to convince the reader that SMRs are a far better option since "compact nuclear reactor systems for surface power represent a significant saving in the launch cost and operate continuously, independent of the sun, for more than 10 years without refueling and with no or little maintenance." The SMRs advocated by these authors, we are assured, are safe if launches do get aborted and units end up being submerged in wet sand or flooded with seawater because of various design features, thus preempting the most obvious argument against the use of nuclear reactors in space.

This vision would see SMRs utilized as early as 2020 to enable the U.S. National Aeronautics and Space Administration (NASA) to achieve its "Vision for Space Exploration," calling for a lunar outpost on the moon by 2020 to serve as the home for 5 to 10 astronauts who would then perform tests on the moon's surface. At a later stage, SMRs would enable the outpost to be expanded to support more elaborate experiments, a greater number of personnel, and the beginning of industrial activity such as the mining of moon minerals and the creation of propellant for future travel to Mars. "Robotic missions to the moon" and to "Mars and beyond," the authors comment, "would require electrical and thermal powers in the order of tens of thousands of kilowatts 24/7, which can be provided using compact and lightweight nuclear reactor power systems as the primary energy source."[106]

Selective remembrance, contradiction, and erasure within SMR visions

Interestingly, there are multiple contradictions, tensions, and tradeoffs inherent in the five SMR visions above, and the use of SMRs to achieve their diverging goals. Even while SMRs are often described as "new" technologies, there is a long history of their development, one occasionally recounted but all too often ignored by proponents. This history, by itself, suggests that similar claims have been made in the past but without these materializing in the form of any tangible benefits. Thus, it is easy to see why this history and these promises are subject to erasure, or "selective remembrance," in the rhetoric surrounding SMRs. Indeed, in my data set, only one paper[107] presented any discussion of the history surrounding the SMR concept.

The concept of having small reactors that can be put together in modules dates back to the first flush of nuclear reactor design and construction. In the 1940s, the United States Army, Air Force, and Navy each initiated research and development programs for various types of small nuclear reactors. The naval program resulted in the light water reactors that powered nuclear submarines and aircraft carriers. The Air Force's Aircraft Nuclear Propulsion program sought to use a SMR to power long-range bombers, and it ran from 1946 to 1961 involving an investment of over $1 billion before being abandoned.[108] Figure 2.2 features one of its proposed aircrafts, the NB-36H *Peacemaker*. The Army Nuclear Power Program ran between 1954 and 1976 and resulted in the construction of eight reactors before being discontinued due to poor economics. These included six 1–2 MWe pressurized water reactors (PWR), one 10 MWe barge-mounted PWR reactor and one 0.5 MWe gas-cooled reactor (GCR). Some of these were located in the kinds of locations that are

FIGURE 2.2 Artist's depiction of the U.S. Air Force's nuclear powered NB-36H *Peacemaker*

Source: U.S. Department of Defense. Note: An air-to-air view of the NB-36H *Peacemaker* experimental aircraft (s/n 51–5,712) during research and development taking place at the Consolidated Vultee Aircraft Corporation (Convair) manufacturing plant in Fort Worth, Texas. Originally named the "Nuclear Test Aircraft" before being designated the NB-36H, the plane carried a 3 MW, air cooled "Aircraft Shield Test Reactor." The NB-36H flew almost 50 flights for a combined flight time of 215 hours, mostly over Texas and New Mexico, during which the reactor operated 89 hours. Defense officials scrapped the plane in September 1958

proposed as potentially attractive sites even today: Antarctica, Greenland, and remote Army bases.

In addition to these defense plans, there were also a number of civilian SMRs proposed for remote locations that had similar power ratings as current designs. Some of these plans received funding from the U.S. Atomic Energy Commission. These included the Nuclear Development Corporation of America's proposal to construct a 100 MW sodium cooled, heavy water moderated reactor in the town of Chugach, Alaska and a 114 MW organic moderated reactor in the city of Piqua, Ohio.[109] Private vendors even touted SMR-powered cars such as the Ford *Nucleon*, a 1958 concept automobile shown in Figure 2.3 to be fueled by a reactor located in the trunk.

Despite the apparent novelty of the space exploration vision, even it was still prominent during this period. For instance, Senator Clinton P. Anderson from New Mexico declared in 1960 that "the universe is nuclear, and so we must use this energy source if we are to go there." Similarly, Krafft Ehricke, a German V-2 rocket scientist, told the U.S. Joint Committee on Atomic Energy that "the universe is run by nuclear energy. Space will be conquered only by manned nuclear-powered vehicles."[110]

Over the next decade or more, based on claims of improved safety and the incorporation of "extraordinary precautions," there were proposals to construct small power reactors in close proximity to population centers. Such proposals declined following the rejection of an audacious proposal by the utility Consolidated Edison to construct a 1,000 MW pressurized water reactor in Queens, New York.[111] Con Edison followed this up with another unsuccessful proposal to construct an underground reactor—an idea that SMR vendors have started emphasizing after Fukushima—on Welfare Island on the East River.[112] With the period of nuclear construction in the United States known as the "great bandwagon market" of the 1970s, and the exploitation of economies of scale

FIGURE 2.3 Illustration of the 1958 Ford *Nucleon*

Source: Henry Ford Museum, image used under a CC BY-NC-ND 4.0 license. Note: The Nucleon was to use a steam engine fueled by uranium fission similar to those found in nuclear submarines designed by Admiral Rickover

resulting in the deployment of reactor designs with approximately 1,000 MW of power capacity, interest in SMRs waned.

The second wave of interest in SMRs came about in the 1980s, as utilities in the United States stopped ordering nuclear reactors. The U.S. Department of Energy observed "new interest in small and medium size reactors and in more advanced reactor concepts other than those marketed today."[113] In 1983, the IAEA launched a new study on what it termed Small and Medium range Power Reactors (SMPRs) with "the objective of surveying the available designs, examining the major factors influencing the decision-making processes in Developing Countries and thereby arriving at an estimate of the potential market."[114] The IAEA "received descriptions of 23 design concepts" but noted that these "had varying levels of readiness and proveness."[115] As of 1987, 19 of these 23 were reported as having documented "readiness and proveness" and the IAEA concluded that SMRs "could play an important role in the 90s and beyond" and reported that several countries had "expressed their particular interest in these reactors."[116] They did not; with the exception of the VVER-440 that was included in this list, no reactor was ever constructed.

In 1991, the Nuclear Development Committee of the Nuclear Energy Agency (NEA) followed up on this IAEA interest by noting that "there is now considerable interest arising in small reactor types."[117] It observed:

> substantial resources being devoted to development of different SMR concepts throughout the "nuclear" nations of the world. The feature common to all these developments is simplification of the design and the introduction of passive safety features … one of the major advantages of SMRs could be in improving public acceptance of nuclear energy, which must be achieved. If nuclear power in general is to play a significant role in the future … The SMR characteristics of simplicity, size and passive safety make them particularly suitable for developing countries.[118]

While the current literature on SMRs does, at times, mention the first wave of interest in the 1950s, and the eventual move to large reactors due to economies of scale, there is no discussion of how the second wave of enthusiasm about SMRs resulted in absolutely no reactors with such designs being constructed. This erasure implies that there is little critical discussion on the challenges to successful implementation of the technology. Indeed, in her work on nuclearity, Gabrielle Hecht argues that the nuclear system creates a "regimes perceptibility" that generates absences and invisibilities with wide-ranging effects—essentially the industry is constantly trying to reinvent itself.[119]

In their comparative survey of nuclear power programs in several countries, scholars John Byrne and Steve Hoffman have noted that nuclear power has been and continues to be evaluated in the "future tense," that is, in terms of what it will bring rather than what it has already wrought, or what it requires from society to maintain operation, "sweeping away current concerns for future gains."[120] Given this longstanding characteristic of nuclear power, it is perhaps unsurprising that many of the visions for SMRs do have an older pedigree—that is, they have an element of continuity with previous rhetorical visions. There are, however, three key reasons for why these new, post-2000 rhetorical visions have taken on a different "flavor," perhaps in order to appeal to a greater number of people.

One factor distinguishing the "new" SMR vision from the "old" is the urgency of climate change. The dire impacts of an increase in global temperature that is now seen as virtually inescapable in a business-as-usual scenario has led to widespread hope for some technological miracle to deliver the world from its potentially civilization-toppling predicament.[121,122] The catch is that nuclear power has been seen—rightly in my opinion—as expensive, susceptible to catastrophic accidents, and associated with undesirable externalities such as the production of weapons-useable fissile material and long-lived radioactive waste. In this scenario, the multiple rhetorical visions put forth by SMR designers and advocates, understandably, offer a great allure, both to themselves as well as a wider public. What is erased in the rhetoric surrounding SMRs is that many, if not all, of these social, economic, and political difficulties will plague the new technology as well.

A second "new" factor involves the vision of offering electricity to countries where significant sections of the population lack access to electricity currently, embedding them in global "energy poverty."[123] The potential role for nuclear power in these developing countries has become particularly important in the last decade or so as the contest between the industrialized and developing countries over responsibility for emissions and allocation of ecological space has driven climate negotiations to a stalemate.[124] Many have underscored the significance of developing countries controlling their fast-growing emissions through low-carbon sources of energy. Therefore, if nuclear power is to meaningfully mitigate greenhouse gas emissions, then it must expand dramatically in these countries, many of which do not have any nuclear power capacity at the moment. SMR advocates posit that unlike traditional reactors, SMRs are capable of meeting this challenge, and thus are attractive to many.

A third "new" factor has been tremendous gains in renewable energy technology with costs of wind and solar power dropping dramatically in the past few years. Further, countries like the United States have seen a dramatic decrease in the cost of natural gas as a result of increased production through hydraulic fracturing. Research across many cultures and communities indicates that the public overwhelmingly prefers the expansion of renewable energy to nuclear power to reduce carbon emissions[125,126,127] The nuclear industry is therefore in a tight spot and a new multi-faceted vision, such as the one offered by SMRs, may be seen as necessary for its survival.

Despite its "newness," however, SMR discourse is still prone to varying types of selective remembrance and erasure. One form is inherent in the vision of environmental nirvana and the use of SMRs to avoid that fate. SMR advocates, and proponents of nuclear energy in general, portray climate change as the only environmental problem paying little attention to other environmental and ecological concerns. Other environmental costs that either go unmentioned or described as manageable are the environmental impacts of uranium mining and the loss of coastal biodiversity due to brine releases from desalination.

An additional type of erasure involves downplaying, or "selectively presenting," data about the cost and economic competitiveness of SMRs. This has been a problem for nuclear power in general and is likely to be exacerbated in the case of SMRs. The most important component of the cost of generating electricity from nuclear reactors is the cost of constructing the facility itself.[128] Current overnight construction costs for the standard sized reactors (roughly 1,000 MWe) are estimated to be in the range of $3,000 to $7,000 per kW of capacity. An example of how these figures translate into total costs is offered by the twin reactors proposed for Vogtle, Georgia in the United States that are estimated to cost $14 billion.[129] Smaller nuclear reactors are typically more expensive on a per unit cost. The general rule of thumb governing capital costs of production facilities is known as the 0.6 power rule.[130] That is, the capital costs of two plants of size S_1 and S_2 are related as:

$$\frac{K_1}{K_2} = \left(\frac{S_1}{S_2}\right)^{0.6}$$

This implies that, all else being equal, a SMR with a power capacity of 200 MW would be expected to have a construction cost in the range of $5,700 to $13,000 per kW of capacity. SMR proponents contest this by claiming that the new reactor designs are as different as to render the comparison with traditional reactor costs invalid. Further, they argue that even if per unit costs are higher because of diseconomies of scale, these can be compensated by the economic advantages accruing from modular and factory construction, learning from replication, and co-siting of multiple reactors.[131] They also posit that there are non-financial factors, such as siting and grid constraints, and impact on the national industrial system, playing important roles in decision-making.[132]

However, detailed and carefully conducted elicitations showed that even experts drawn from, or closely associated with, the nuclear industry, most of whom worked for major US reactor vendors that are actively developing commercial SMRs, expect these reactors to cost more per kW of capacity than currently operating reactors.[133] In turn, this higher cost per kW of capacity translates into a higher cost per kWh of electrical energy—undermining the investment potential, profitability, and social affordability and acceptability of SMRs.[134,135]

Given these economic challenges, the likelihood of SMRs being able to meet many of the visions, in particular indigenous self-energization, is low. The main obstacle that confronts governments and private companies in trying to provide electricity for the hundreds of millions of people who do not have access to it currently is the fact that most of them cannot afford to purchase electricity. Low cost is, therefore, a vital consideration in trying to design electricity sources to meet their needs. The expected high cost per kWh implies that SMR generated electricity will likely be unaffordable to the target population touted in the indigenous self-energization vision.

The cost problem is exacerbated when the technical requirements for meeting other visions are appended. For example, the strategy advocated by the publication that postulated SMRs as a viable route for the "elimination of long-lived radioactive wastes" is through the use of the transmutation of radionuclides with long half-lives. There are technical problems with transmutation, and it is not possible to entirely eliminate all troublesome isotopes. More to the point, however, is the enormous cost of this approach; the U.S. National Academy of Sciences estimated that the additional cost of transmuting the nuclear waste in the United States was "likely to be no less than $50 billion and easily could be over $100 billion."[136]

Likewise, efforts to lower the risk of radioactivity release to the atmosphere from a reactor accident by constructing SMRs underground would result in a higher construction cost. The absence of widespread discussion of the cost implications of underground siting is another case of erasure. The idea of underground siting was studied extensively in the 1970s and early 1980s, but these studies concluded there would be an almost certain cost increase due to the need to excavate large volumes of rock and the additional complications of constructing a complex facility below the ground; for example, a study by Ontario Hydro from Canada showed a 31–36 percent increase relative to an equivalent surface facility, and warned about "potentially more difficult construction and operating procedures."[137]

In other words, the multiple visions put forth about SMRs suffer from contradictions. The literature on SMRs is typically silent about these contradictions, and erases the fact that no single design can simultaneously meet the technical requirements of all the rhetorical visions associated with SMRs. For example, SMRs based on LWRs, the kind that are expected to be first constructed in the United States, would use more uranium per unit of electrical energy generated (kWh or MWh) than standard-sized nuclear reactors.[138] This implies also that a greater quantity of radioactive spent fuel will be generated for the same amount of energy, and that, all else being equal, there would be a greater risk of weapons proliferation and environmental degradation associated with the facility. Other SMR designs with long-lived cores that would reduce the quantities of radioactive waste generated have a higher fissile (plutonium) content in the spent fuel they generate, again increasing the associated risk of proliferation.

Conclusion

The nuclear industry and its attendant institutions, including sections of academia and the IAEA, have created a number of rhetorical visions related to SMRs, in turn propelling a technological utopianism among promoters, political leaders, and financial investors. What is perhaps surprising is that practically all of the articles in my sample are about the abstruse technical details of SMRs, and yet they feature rhetorical visions promising a future replete with risk-free reactors, generating clean electricity propelling economic growth, universal access to drinking water, a more sustainable climate and environment, and the establishment of outposts on the Moon, Mars, and "beyond." Underlying these specific visions is a more general utopian fantasy, a society that requires and provides increasing and abundant quantities of energy indefinitely into the future. One scholar has termed this sort of fantasy "the dream of a *perpetuum mobile.*"[139]

The term "fantasy" here is not meant to imply that those scientists sharing it are collectively neurotic. The types of fantasies associated with SMRs satisfy the human need to experience and interpret drama and enable people to feel positive about the future. They may also be ways of diverting attention from the many problems confronting commercialization of the dozens of SMR concepts that have been put forward for some decades now. Though each of these rhetorical visions had varying supporters, arguments, narratives, symbolic cues, audiences, functions, characters, and degrees of frequency, I believe their presence nonetheless has broader implications for SMR advocates, for the nuclear renaissance, for the practice of science, and for energy policymaking.

First, for SMR advocates, my study reveals that there is no common, single vision of what SMRs can accomplish. Instead, as predicted by the literature, such visions are full of contradictions, erasures, and tension. There are competing, at times overlapping visions, not all of them consistent, some of them part of larger visions that cut across a variety of nuclear technologies. For example, the vision of indigenous self-energization would see SMRs placed in all communities that desire economic growth (conceivably most of them) regardless of its environmental effects; the environmental vision utilizes SMRs to prioritize preservation of the environment and the earth's climatic system.

Furthermore, many individual documents from the sample had multiple themes, especially longer IAEA reports and conference proceedings (some as long as 712 pages) which blended together rhetorical visions. For example, one IAEA report fused together concerns about climate change, the environment, and water desalination when it argued that:

Desalination with fossil energy sources would not be compatible with sustainable development: fossil fuels reserves are finite and must be conserved for other essential uses whereas demands for desalted water would continue to increase. Furthermore, the combustion of fossil fuels would produce large amounts of greenhouse gases and toxic emissions, underlining the need for options involving alternative energy sources [such as SMR reactors].[140]

Other studies hybridized separate rhetorical visions into something new, or gave their visions a distinct cultural twist. In Mongolia, SMR proponents merge together the environment and self-energization themes when they argue that "using nuclear energy can be one of the ways to satisfy increasing energy demand and to solve the air pollution issues in Mongolia."[141] In Russia, widespread uses of SMRs for electric heating is depicted as ensuring "fossil fuel and power-independence" (a vision of self-energization) as well as "replacement of fossil fuel power sources" that have become obsolete due to environmental concerns (a vision of environmental nirvana).[142] In the United States, SMRs are sold for their "excellent safety and performance record" (a vision of risk-free energy) and a growing concern for the "environmental impacts of fossil fuels" (a vision of environmental nirvana).[143]

These examples point to the inherently flexible nature of rhetorical visions, perhaps strategically so. They indicate that some visions contradict others, or may argue that particular countries can benefit at the exclusion of others. They confirm that no single rhetorical vision is universally persuasive, or subscribed to by a solid majority of institutions. And rather than carefully or systematically analyzing the promise and perils of SMRs, most proponents view them merely as a platform to advance their own agendas. The complex history behind SMRs is thus "erased" in exchange for more narrow or powerful narratives that serve the vested interests of particular stakeholders.

Another conclusion is that, for the nuclear renaissance, the visions identified with SMRs may accentuate why nuclear power has so much appeal for policymakers and the mass public. It suggests that any challenges for next generation reactors, including SMRs, may be severely discounted in the wake of the much more powerful and compelling fantasies associated with what nuclear technology can someday accomplish. SMRs are particularly endearing because there are so many different designs available to satisfy the public's imagination. Moreover, statements about an SMR future are overly deterministic by presenting the future as a fixed and relatively stable extension of current events. This has the unintended, or perhaps even intended, effect of disguising a degree of social choice in energy planning, presenting a SMR future as knowable and inevitable, rather than the result of strategic decisions and social practices. The fantasy thus becomes even more compelling since it involves little to no sacrifice, and minimal effort on the part of the public.[144]

Additionally, for scientific (and engineering) practice, my study affirms that scientists and engineers are not immune to the powers of drama and fantasy, and that they can become "infected" with technological utopianism that cause them, in their excitement, to lose their scientific precision. This may be relatively unique to physicists—Brysse et al. for instance noted that atmospheric and climate scientists tried to edge the other way, on the side of the "least drama" or the least alarming visions.[145] Nuclear physicists, on the other hand, headed the opposite way. Most of the sampled documents were exceedingly technical, yet were full of unscientific language indicating, for instance, that accidents can be "solved" and "eliminated" rather than the more accurate terminology that their probabilities can be lowered or that safety can be "enhanced" or "improved." The claims made were very reminiscent of the early years of the nuclear age, which later had to be "explained" away.

Deviations within the sample of SMR visions also occur from historical utopias of the past. There are elements of Segal's evolutionary, efficient, and anthropocentric utopias: SMRs will need investment and innovation before they can become risk free; they are lauded for their ability to produce energy with minimal impact on the environment, or by using local and indigenous knowledge and resources; and they are hoped to eventually create a world of peace and plenty where humans can thrive or are even empowered to explore other extraterrestrial worlds. But I do not see SMR visions that touch upon rational utopian themes about conquering emotion, individualistic utopian themes about enhancing self-status or awareness, or imperial utopian themes about conquering other peoples or cultures. Perhaps this implies such visions as a whole have moved beyond the "cognitive imperialism" feared by Marvin and envision a more pluralistic (though admittedly still anthropocentric) future.

Lastly, there is an inherent tension in many of the claims made by SMR advocates about innovation, learning, and the adaptability of designs and the process of erasure. Simply put: how can the industry learn from the past if it erases its own history? Also, why isn't the institutional memory of an organization such as the IAEA stronger? At an even deeper and potentially more worrying level, the conditions of erasure under which SMR programs operate complicates the process of obtaining data for a comprehensive cost analysis of nuclear power. These conditions make it impossible for consumers of nuclear electricity (and policymakers) to make "rational" and informed choices about electricity supply. Moreover, the process of erasure has troubling implications for nuclear safety, as it by definition obfuscates historical facts. Consequently, both public and political support for SMRs remains rooted in an incomplete knowledge set—health threats, environmental threats and program costs are all implicated.

APPENDIX 2.1 Details of my final sample of 60 articles depicting SMR visions

	Reference	First author institution	First author country
Academic studies	(Bortot et al. 2011)[146]	Politecnico di Milano, Department of Energy, CeSNEF-Nuclear Engineering Division	Italy
	(Carelli et al. 2004)[147]	Westinghouse Electric Company	United States
	(Carelli et al. 2010)[148]	Westinghouse Electric Company	United States
	(Choi et al. 2011)[149]	Department of Nuclear Engineering, Seoul National University	South Korea
	(Chung et al. 2004)[150]	Korea Atomic Energy Research Institute	South Korea
	(Ding and Kloosterman 2011)[151]	Delft University of Technology	Netherlands
	(El-Genk and Tournier 2004)[152]	Institute for Space and Nuclear Power Studies, University of New Mexico	United States
	(Filho 2011)[153]	Instituto de Engenharia Nuclear (IEN/ CNEN)	Brazil
	(Fuls and Mathews 2007)[154]	North West University, CRCED Pretoria	South Africa
	(Hatton and El-Genk 2009)[155]	Institute for Space and Nuclear Power Studies, University of New Mexico	United States
	(Hibi, Ono, and Kanagawa 2004)[156]	Mitsubishi Heavy Industries	Japan
	(Hong, Kim, and Venneri 2008)[157]	Korea Atomic Energy Research Institute	South Korea
	(Ingersoll 2009)[158]	Oak Ridge National Laboratory	United States
	(Kang, Tak, and Kim 2012)[159]	Korea Atomic Energy Research Institute	South Korea
	(Karahan 2010)[160]	Nuclear Science and Engineering Department, Massachusetts Institute of Technology	United States
	(Kessides 2012)[161]	The World Bank, Development Research Group	United States
	(Lee, Kim, and Park 2012)[162]	Department of Nuclear Engineering, Seoul National University	South Korea
	(Lim et al. 2011)[163]	Department of Mechanical Engineering, Yonsei University	South Korea
	(Matsumura et al. 2005)[164]	Central Research Institute of Electric Power Industry	Japan
	(Şahin and Sefidvash 2008)[165]	Nuclear Engineering Department, Federal University of Rio Grande do Sul	Brazil

(Continued)

	Reference	First author institution	First author country
	(Sambuu and Obara 2012)[166]	Nuclear Research Center, National University of Mongolia	Mongolia
	(Shropshire 2011)[167]	European Commission, Joint Research Centre	Netherlands
	(Slessarev 2007)[168]	N/A	France
	(Slessarev 2008)[169]	N/A	France
	(Tominaga et al. 2005)[170]	Hitachi Power Systems Limited	Japan
	(Toshinsky, Komlev, and Mel'nikov 2011)[171]	State Scientific Center of Russian Federation, Institute for Physics and Power Engineering	Russia
	(Vujić et al. 2012)[172]	University of California Berkeley	United States
	(Wallace et al. 2006)[173]	Pebble Bed Modular Reactor Limited	South Africa
	(Yang et al. 2006)[174]	Fluid System Engineering Department, Korea Atomic Energy Research Institute	South Korea
	(Yoon et al. 2012)[175]	Department of Nuclear and Quantum Engineering, Korea Advanced Institute of Science and Technology	South Korea
	(Zhang and Sun 2007)[176]	Institute of Nuclear and New Energy Technology (INET), Tsinghua University, Beijing	China
	(Zheng et al. 2012)[177]	Institute of Nuclear and New Energy Technology (INET), Tsinghua University, Beijing	China
	(Zrodnikov et al. 2008)[178]	FSUE State Scientific Center of Russian Federation Institute for Physics and Power Engineering	Russia
	(Zrodnikov et al. 2011)[179]	FSUE State Scientific Center of Russian Federation Institute for Physics and Power Engineering	Russia
IAEA Publications	International Atomic Energy Agency	*Methodology for the Assessment of Innovative Nuclear Reactors and Fuel Cycles* (Vienna: International Atomic Energy Agency, December 2004, IAEA-TECDOC-1434)	
	International Atomic Energy Agency	*Innovative small and medium sized reactors: Design features, safety approaches and R&D trends* (Vienna: IAEA, May, 2005, IAEA-TECDOC-1451)	
	International Atomic Energy Agency	*Status of innovative small and medium sized reactor designs 2005: Reactors with conventional refuelling schemes* (Vienna: IAEA, March, 2006, IAEA-TECDOC-1485)	

(*Continued*)

Reference	First author institution	First author country
International Atomic Energy Agency	*Advanced nuclear plant design options to cope with external events* (Vienna: IAEA, February, 2006, IAEA-TECDOC-1487)	
International Atomic Energy Agency	*Nuclear Technology Review 2007* (Vienna, Austria: IAEA, 2007a)	
International Atomic Energy Agency	*Advanced Applications of Water Cooled Nuclear Power Plants* (Vienna: IAEA, IAEA-TECDOC-1584, 2007b).	
International Atomic Energy Agency	*Status of Small Reactor Designs Without On-Site Refuelling* (Vienna: International Atomic Energy Agency, IAEA-TECDOC-1536, 2007c).	
International Atomic Energy Agency	*Non-Electric Applications of Nuclear Power: Seawater Desalination, Hydrogen Production, and Other Industrial Applications* (Vienna: IAEA, IAEA-CN-152, 2009a)	
International Atomic Energy Agency	*Common User Considerations (CUC) by Developing Countries for Future Nuclear Energy Systems: Report of Stage 1* (Vienna: IAEA, No. NP-T-2.1, 2009b)	
International Atomic Energy Agency	*Design Features to Achieve Defence in Depth in Small and Medium Sized Reactors* (Vienna: IAEA, No. NP-T-2.2, 2009c).	
International Atomic Energy Agency	*Passive Safety Systems and Natural Circulation in Water Cooled Nuclear Power Plants* (Vienna: International Atomic Energy Agency, IAEA-TECDOC-1624, 2009d)	
International Atomic Energy Agency	*Lessons Learned from Nuclear Energy System Assessments (NESA) Using the INPRO Methodology. A Report of the International Project on Innovative Nuclear Reactors and Fuel Cycles (INPRO)* (Vienna: International Atomic Energy Agency, IAEA-TECDOC-1636, 2009e)	
International Atomic Energy Agency	*Status and Trends of Nuclear Technologies Report of the International Project on Innovative Nuclear Reactors and Fuel Cycles (INPRO)* (Vienna: International Atomic Energy Agency, IAEA-TECDOC-1622, 2009)	
International Atomic Energy Agency	*Environmental Impact Assessment of Nuclear Desalination* (Vienna: IAEA, IAEA-TECDOC-1642, March 2010a)	
International Atomic Energy Agency	*Small Reactors Without On-site Refueling: Neutronic Characteristics, Emergency Planning, and Development Scenarios* (Vienna: IAEA, September, IAEA-TECDOC-1652, 2010b	

(Continued)

Reference	First author institution	First author country
International Atomic Energy Agency	*Assessment of Nuclear Energy Systems Based on a Closed Nuclear Fuel Cycle with Fast Reactors: A Report of the International Project on Innovative Nuclear Reactors and Fuel Cycles (INPRO)* (Vienna: International Atomic Energy Agency, IAEA-TECDOC-1639, 2010)	
International Atomic Energy Agency	*Status of Small and Medium Sized Reactor Designs* (Vienna: IAEA, Nuclear Power Technology Development Section Division of Nuclear Power—Department of Nuclear Energy, September, 2011).	
International Atomic Energy Agency	*Fast Reactors and Related Fuel Cycles: Challenges and Opportunities* (Vienna: IAEA FR09, March, 2012a, STI/PUB/1444)	
International Atomic Energy Agency	*Advances in Nuclear Power Process Heat Applications* (Vienna: IAEA, IAEA-TECDOC-1682, 2012b)	
International Atomic Energy Agency	*Natural Circulation Phenomena and Modelling for Advanced Water Cooled Reactors* (Vienna: International Atomic Energy Agency, IAEA-TECDOC-1677, 2012c)	
International Atomic Energy Agency	*Advances in High Temperature Gas Cooled Reactor Fuel Technology* (Vienna: International Atomic Energy Agency, IAEA-TECDOC-1674, 2012)	
Kuznetsov	"Design Status and Applications of Small Reactors Without On-Site Refuelling," Proceedings of the 14th International Conference on Nuclear Engineering ICONE 16, 17–20 July 2006, Orlando, Florida, USA, Paper ICONE 14–89318	
Kuznetsov	"Advanced Nuclear Plant Design Options to Cope with External Events," Proceedings of the 19th International Conference on Structural Mechanics in Reactor Technology SMiRT 19, Toronto, Canada, 12–17 August 2007, Paper S01/2	
Kuznetsov	"Design and Technology Development Status and Design Considerations for Innovative Small and Medium Sized Reactors," Proceedings of the 16th International Conference on Nuclear Engineering ICONE 16, 11–15 May 2008, Orlando, Florida, USA, Paper ICONE16-48048	

(Continued)

APPENDIX 2.1 (Cont.)

Reference	First author institution	First author country
Kuznetsov	"Design Features to Achieve Defence in Depth in Small and Medium Sized Reactors," Proceedings of the International Conference on Fast Reactors and Related Fuel Cycles: Challenges and Opportunities, Kyoto, Japan, 7–11 December 2009, Paper IAEA-CN-176	
Kuznetsov et al	"Approaches to Assess Competitiveness of Small and Medium Sized Reactors," Proceedings of the 17th International Conference on Nuclear Engineering ICONE 17, 12–16 July 2009, Brussels, Belgium, Paper ICONE17-75597	

Notes

1 Bocking, S. 2009, February 15. Dams as Development. Globalization Monitor. www.globalmon.org.hk/content/dams-development.
2 Ibid.
3 Nye, D. E. 1994. American Technological Sublime (Cambridge, MA: MIT Press).
4 Khagram, S. 2005. Beyond Temples and Tombs: Towards Effective Governance for Sustainable Development through the World Commission on Dams (Cambridge, MA: Center for International Development and Hauser Center for Non-Profit Organizations, The John F. Kennedy School of Government, Harvard University).
5 Sovacool, B. K., and M. V. Ramana. 2015, January. Back to the Future: Small Modular Reactors, Nuclear Fantasies, and Symbolic Convergence. Science, Technology, & Human Values 40 (1): 96–125.
6 Ingersoll, D. T. 2009. Deliberately Small Reactors and the Second Nuclear Era. Progress in Nuclear Energy 51 (4–5): 589–603.
7 Kessides, I. N. 2012. The Future of the Nuclear Industry Reconsidered: Risks, Uncertainties, and Continued Promise. Energy Policy 48: 185–208.
8 Siegrist, M., and V. H. M. Visschers. 2013. Acceptance of Nuclear Power: The Fukushima Effect. Energy Policy 59: 112–119.
9 Sovacool, B., and V. Scott Victor. 2012. The National Politics of Nuclear Power: Economics, Security, and Governance (London; New York: Routledge).
10 Ingersoll, D. T. 2016. Energy, Nuclear Power, and Small Modular Reactors. In Small Modular Reactors Nuclear Power Fad or Future? (New York: Woodhead Publishing), pp. 3–19.
11 Matthews, J. 2015, November 18. Small Modular Reactors—The Real Nuclear Renaissance? University of Manchester, Featured, Science and Technology.
12 Johnston, H. 2018, January 26. Small Modular Nuclear Reactors are a Crucial Technology, Says Report. Physics World.
13 IEA. 2013. RD&D Budgets of OECD Countries. Energy Technology RD&D, p.2. http://wds.iea.org/wds/ReportFolders/ReportFolders.aspx.
14 Cowan, R. 1990. Nuclear Power Reactors: A Study in Technological Lock-In. The Journal of Economic History 50 (3): 541–567.
15 Selin, C. 2007. Expectations and the Emergence of Nanotechnology. Science, Technology & Human Values 32 (2): 196–220.

16 Nuclear Power Reactors in the World: 2011 Edition (Vienna: International Atomic Energy Agency).
17 International Atomic Energy Agency. 2014, September. Updated Status on Global SMR Development (Vienna: IAEA).
18 Noble, D. F. 1997. The Religion of Technology: The Divinity of Man and the Spirit of Invention (New York: Penguin Books), p. 208.
19 Berkout, F. 2006, July–September. Normative Expectations in Systems Innovation. Technology Analysis & Strategic Management 8 (3/4): 299–311.
20 Segal, H. P. 2005. Technological Utopianism in American Culture (Syracuse, NY: Syracuse University Press).
21 Segal, H. P. 1994. Future Imperfect: The Mixed Blessings of Technology in America (Boston, MA: University of Massachusetts).
22 Halley, R. B., and H. G. Vatter. 1978, January. Technology and the Future as History: A Critical Review of Futurism. Technology and Culture 19 (1): 53–82.
23 Segal (2005).
24 Marvin, C. 1988. When Old Technologies Were New: Thinking about Electric Communication in the Late Nineteenth Century (Oxford: Oxford University Press).
25 Rejeski, D., and R. L. Olson. 2006, Winter. Has Futurism Failed? Wilson Quarterly 14: 8.
26 Eisenstaedt, S. N. 1982, Watersheds. The Axial Age: The Emergence of Transcendental Visions and the Rise of Clerics. European Journal of Sociology/Archives Européennes De Sociologie/Europäisches Archiv Für Soziologie 23 (2): 294–314.
27 Corn, J. J. 1986. Imagining Tomorrow: History, Technology and the American Future (Cambridge, MA: MIT Press).
28 Sturken, M., D. Thomas, and S. J. Ball-Rokeach (Eds.). 2004. Technological Visions: The Hopes and Fears that Shape New Technologies (Philadelphia, PA: Temple University Press).
29 Marvin (1988).
30 Ibid., 192.
31 Ibid.
32 Editors of Pocket Books. 1945. The Atomic Age Opens (New York: Pocket Books), pp. 202–203.
33 Makhijani, A., and S. Saleska. 1999. The Nuclear Power Deception: U.S. Nuclear Mythology from Electricity Too Cheap to Meter to Inherently Safe Reactors (New York: Apex Press), p. 17.
34 Ibid., 54.
35 Ford, D. 1986. Meltdown: The Secret Papers of the Atomic Energy Commission (New York: Simon & Shuster), p. 30.
36 Strauss, L. 1954, September 17. Speech to the National Association of Science Writers, September 16th, 1954. New York Times, p. 1A.
37 Asimov, I. 1964, August 16. Visit to the World's Fair of 2014. New York Times.
38 Cohn, S. M. 1997. Too Cheap to Meter: An Economic and Philosophical Analysis of the Nuclear Dream (New York: State University of New York Press), p. 19.
39 Lilienthal, D. E. 1963. Change, Hope, and the Bomb (Princeton, NJ: Princeton University Press), pp. 109–110.
40 Sovacool and Ramana (2015).
41 McDowall, W., and M. Eames. 2006. Forecasts, Scenarios, Visions, Backcasts and Roadmaps to the Hydrogen Economy: A Review of the Hydrogen Futures Literature. Energy Policy 34 (11): 1236–1250.
42 Sovacool, B. K., and B. Brossmann. 2010. Symbolic Convergence and the Hydrogen Economy. Energy Policy 38 (4): 1999–2012.
43 International Atomic Energy Agency. About Us. Accessed 5 September 2012, www.iaea.org/.

44 As the IAEA Summarized in 2009, Recently, More than 50 Concepts and Designs of Such Innovative SMRs Were Developed in Argentina, Brazil, China, France, India, Japan, the Republic of Korea, Russian Federation, South Africa, and the USA (IAEA 2009a: 2).

45 Sovacool and Ramana (2015: 96–125).

46 Mourogov, V., K. Fukuda, and V. Kagramanian. 2002. The Need for Innovative Nuclear Reactor and Fuel Cycle Systems: Strategy for Development and Future Prospects. Progress in Nuclear Energy 40 (3–4): 286.

47 Ibid., 285–299.

48 Sovacool and Ramana (2015: 96–125).

49 IAEA. 2006a. Advanced Nuclear Power Plant Design Options to Cope with External Events. IAEA-TECDOC-1487 (Vienna: International Atomic Energy Agency).

50 Slessarev, I. 2007. Intrinsically Secure Fast Reactors with Dense Cores. Annals of Nuclear Energy 34 (11): 883–895.

51 Filho, G., and O. J. Agostinho. 2011. INPRO Economic Assessment of the IRIS Nuclear Reactor for Deployment in Brazil. Nuclear Engineering and Design 241 (6): 2329–2338.

52 Zrodnikov, A. V. et al. 2008. Innovative Nuclear Technology Based on Modular Multi-Purpose Lead–Bismuth Cooled Fast Reactors. Progress in Nuclear Energy 50 (2–6): 170–178.

53 IAEA. 2007a. Nuclear Technology Review 2007 (Vienna: International Atomic Energy Agency).

54 IAEA. 2009. Design Features to Achieve Defence in Depth in Small and Medium Sized Reactors. IAEA-NP-T-2.2 (Vienna: International Atomic Energy Agency).

55 Zhang, Z., and Y. Sun. 2007. Economic Potential of Modular Reactor Nuclear Power Plants Based on the Chinese HTR-PM Project. Nuclear Engineering and Design 237 (23): 2265–2274.

56 Zrodnikov, A. V. et al. 2011. SVBR-100 Module-Type Fast Reactor of the IV Generation for Regional Power Industry. Journal of Nuclear Materials 415 (3): 237–244.

57 Hibi, K., H. Ono, and T. Kanagawa. 2004. Integrated Modular Water Reactor (IMR) Design. Nuclear Engineering and Design 230 (1–3): 253–266.

58 Wallace, E. et al. 2006. From Field to factory—Taking Advantage of Shop Manufacturing for the Pebble Bed Modular Reactor. Nuclear Engineering and Design 236: 445–453.

59 Yang, S. H. et al. 2006. Overpressure Protection Analysis of an Advanced Integral Reactor. Nuclear Engineering and Design 236 (22): 2376–2385.

60 Ingersoll (2009: 592).

61 El-Genk, M. S., and J.-M. P. Tournier. 2004. AMTEC/TE Static Converters for High Energy Utilization, Small Nuclear Power Plants. Energy Conversion and Management 45 (4): 511–535.

62 IAEA (2006a).

63 Choi, S. et al. 2011. PASCAR: Long Burning Small Modular Reactor Based on Natural Circulation. Nuclear Engineering and Design 241 (5): 1486–1499.

64 IAEA (2007a: 93).

65 Ibid., 93–94.

66 IAEA. 2009. Common User Considerations (CUC) by Developing Countries for Future Nuclear Energy Systems: Report of Stage 1. NP-T-2.1 (Vienna: International Atomic Energy Agency).

67 IAEA. 2005. Innovative Small and Medium Sized Reactors: Design Features, Safety Approaches and R&D Trends. IAEA-TECDOC-1451 (Vienna: International Atomic Energy Agency).

68 Kuznetsov, V. 2008. Design and Technology Development Status and Design Considerations for Innovative Small and Medium Sized Reactors. In 16th International Conference on Nuclear Engineering. Orlando, FL.

69 IAEA. 2007. Status of Small Reactor Designs without On-Site Refuelling. Nuclear Energy Series IAEA-TECDOC-1536 (Vienna: International Atomic Energy Agency).

70 Vujić, J. et al. 2012. Small Modular Reactors: Simpler, Safer, Cheaper? Energy 45 (1): 288–295.

71 El-Genk and Tournier (2004: 511–512).

72 Ingersoll (2009: 602).

73 Zrodnikov et al. (2011: 24).

74 Tominaga, K. et al. 2005. Development of Medium and Small Sized Reactors: DMS. Progress in Nuclear Energy 47 (1–4): 106–114.

75 Bortot, S. et al. 2011. Core Design Investigation for a SUPERSTAR Small Modular Lead-Cooled Fast Reactor Demonstrator. Nuclear Engineering and Design 241 (8): 3021–3031.

76 IAEA (2007b: 183).

77 Ibid., 479.

78 IAEA. 2004. Methodology for the Assessment of Innovative Nuclear Reactors and Fuel Cycles. IAEA-TECDOC-1434 (Vienna: International Atomic Energy Agency).

79 Slessarev, I. 2008. Intrinsically Secure Fast Reactors for Long-Lived Waste Free and Proliferation Resistant Nuclear Power. Annals of Nuclear Energy 35 (4): 636.

80 Kessides (2012: 187).

81 IAEA (2006b: 1).

82 IAEA (2009a: V).

83 IAEA. 2012. Fast Reactors and Related Fuel Cycles: Challenges and Opportunities. STI/PUB/1444 (Vienna: International Atomic Energy Agency).

84 Vujić et al. (2012: 288).

85 Shropshire, D. 2011. Economic Viability of Small to Medium-Sized Reactors Deployed in Future European Energy Markets. Progress in Nuclear Energy 53 (4): 299–307.

86 Slessarev (2008: 637).

87 IAEA (2010a: 6).

88 IAEA (2007c: 34).

89 Ibid., 33–34).

90 IAEA (2009a: 2).

91 El-Genk and Tournier (2004: 512).

92 IAEA (2010b: 3).

93 IAEA (2012b).

94 IAEA (2010b: 58–60).

95 Ibid., 57.

96 Ibid., 59.

97 Ingersoll (2009: 598).

98 IAEA (2011: 1).

99 Lim, S. et al. 2011. Dynamic Characteristics of a Perforated Cylindrical Shell for Flow Distribution in SMART. Nuclear Engineering and Design 241 (10): 4079–4088.

100 IAEA (2005: 89–96).

101 IAEA (2006b: 64).

102 It was articulated in only two studies within our sample, although there have been significant efforts over the decades invested in developing space nuclear reactor systems and numerous papers do show up on ScienceDirect, except that they don't usually use SMR as a keyword.

103 Hatton, S. A., and M. S. El-Genk. 2009. Sectored Compact Space Reactor (Score) Concepts with a Supplementary Lunar Regolith Reflector. Progress in Nuclear Energy 51 (1): 93–108.

104 IAEA (2007b: 469).

105 Hatton and El-Genk (2009: 93).

106 Ibid.

107 Ingersoll (2009).

108 Ibid., 590.

109 Department of Energy. 1958. Quarterly Progress Report (Washington, DC: Department of Energy).

110 Mahaffey, J. 2010. Atomic Awakening: A New Look at the History and Future of Nuclear Power (New York: Pegasus), p. 277.

111 Buckley, T. 1963, May 10. City to Weigh Peril of Nuclear Plant Sought for Queens. New York Times, p.2.

112 Bird, D. 1968, October 7. Nuclear Plant Proposed Beneath Welfare Island. New York Times, p.4.

113 Department of Energy. 1988. Nuclear Energy Cost Data Base. DOE/NE-0095. Office of Program Support, U.S. Department of Energy. www.ne.doe.gov/energy PolicyAct2005/neEPACT2a.html.

114 IAEA. 1985. Small and Medium Power Reactors, Project Initiation Study, Phase 1. IAEA-TECDOC-347 (Vienna: International Atomic Energy Agency).

115 Konstantinov, L. V., and J. Kupitz. 1988. The Status of Development of Small and Medium Sized Reactors. Nuclear Engineering and Design 109 (1–2): 5–9.

116 Ibid., 9.

117 NEA. 1991. Small and Medium Reactors: Status and Prospects (Paris: Nuclear Energy Agency, OECD).

118 Ibid., 100.

119 Hecht, G. 2012. Being Nuclear: Africans and the Global Uranium Trade (Cambridge, MA: MIT Press).

120 Byrne, J., and S. M. Hoffman. 1996. The Ideology of Progress and the Globalization of Nuclear Power. In Byrne, J. and Hoffman, S. M. (Eds.): Governing the Atom: The Politics of Risk (New Brunswick: Transaction Publishers), pp. 11–46.

121 Rockstrom, J. et al. 2017. A Roadmap for Rapid Decarbonisation. Science 355: 1269–1271.

122 Figueres, C. et al. 2017. Three Years to Safeguard Our Climate. Nature 546: 593–595.

123 Halff, A., J. Rozhon, and B. K. Sovacool (Eds.). 2014. Energy Poverty: Global Challenges and Local Solutions (Oxford: Oxford University Press).

124 Dubash, N. K. (Ed.). 2012. Handbook of Climate Change and India: Development, Politics and Governance (New York: Earthscan).

125 Pidgeon, N. F., I. Lorenzoni, and W. Poortinga. 2008. Climate Change or Nuclear power–No Thanks! A Quantitative Study of Public Perceptions and Risk Framing in Britain. Global Environmental Change 18 (1): 69–85.

126 Greenberg, M. 2009. Energy Sources, Public Policy, and Public Preferences: Analysis of US National and Site-Specific Data. Energy Policy 37 (8): 3242–3249.

127 Ertör-Akyazı, P. et al. 2012. Citizens' Preferences on Nuclear and Renewable Energy Sources: Evidence from Turkey. Energy Policy 47: 309–320.

128 Ramana, M. V. 2009. Nuclear Power: Economic, Safety, Health, and Environmental Issues of Near-Term Technologies. Annual Review of Environment and Resources 34: 127–152.

129 Smith, R. 2012, February 8. WSJ UPDATE: Southern Co. Near Approvals for 2 Reactors. Wall Street Journal. http://online.wsj.com/article/BT-CO-20120208-718884.html.

130 National Research Council. 1996. Nuclear Wastes: Technologies for Separations and Transmutation (Washington, DC: National Academy Press).

131 Carelli, M. et al. 2010. Economic Features of Integral, Modular, Small-To-Medium Size Reactors. Progress in Nuclear Energy 52 (4): 403–414.

132 Locatelli, G., and M. Mancini. 2011, March. The Role of the Reactor Size for an Investment in the Nuclear Sector: An Evaluation of Not-Financial Parameters. Progress in Nuclear Energy 53 (2): 212–222.

133 Abdulla, A., I. L. Azevedo, and M. Granger Morgan. 2013, June. Expert Assessments of the Cost of Light Water Small Modular Reactors. Proceedings of the National Academy of Sciences 110 (24): 9686–9691.

134 Ramana, M. V., and Z. Mian. 2014, June. One Size Doesn't Fit All: Social Priorities and Technical Conflicts for Small Modular Reactors. Energy Research & Social Science 2: 115–124.

135 Cooper, M. 2014, September. Small Modular Reactors and the Future of Nuclear Power in the United States. Energy Research & Social Science 3: 161–177.

136 National Research Council (1996: 7).

137 Oberth, R. C., and C. F. Lee. 1980. Underground Siting of a CANDU Nuclear Power Station. In Rockstore 80 (Stockholm, Sweden: Pergamon Press).

138 Glaser, A., L. B. Hopkins, and M. V. Ramana. 2013. Resource Requirements and Proliferation Risks Associated with Small Modular Reactors. Nuclear Technology 184 (1): 121–129.

139 Hultman, M. 2009. Back to the Future: The Dream of a Perpetuum Mobile in the Atomic Society and the Hydrogen Economy. Futures 41 (4): 226–233.

140 IAEA (2010b: 2).

141 Sambuu, O., and T. Obara. 2012. Conceptual Design for a Small Modular District Heating Reactor for Mongolia. Annals of Nuclear Energy 47: 210–215.

142 Zrodnikov et al. (2011).

143 Ingersoll (2009).

144 Walker, M. 2006. America's Romance with the Future. The Wilson Quarterly 30 (1): 22–26.

145 Brysse, K. et al. 2013. Climate Change Prediction: Erring on the Side of Least Drama? Global Environmental Change 23: 327–337.

146 Bortot et al. (2011: 3021–3031).

147 Carelli, M. et al. 2004. The Design and Safety Features of the IRIS Reactor. Nuclear Engineering and Design 230 (1–3): 151–167.

148 Carelli et al. (2010: 403–414).

149 Choi et al. (2011: 1486–1499).

150 Chung, Y.-J. et al. 2004. Thermal Hydraulic Calculation in a Passive Residual Heat Removal System of the SMART-P Plant for Forced and Natural Convection Conditions. Nuclear Engineering and Design 232 (3): 277–288.

151 Ding, M., and J. L. Kloosterman. 2011. Neutronic Feasibility Design of a Small Long-Life HTR. Nuclear Engineering and Design 241 (12): 5093–5103.

152 El-Genk and Tournier (2004: 511–535).

153 Filho and Agostinho (2011: 2329–2338).

154 Fuls, W. F., and E. H. Mathews. 2007. Passive Cooling of the PBMR Spent and Used Fuel Tanks. Nuclear Engineering and Design 237 (12–13): 1354–1362.

155 Hatton and El-Genk (2009: 93–108).

156 Hibi, Ono, and Kanagawa (2004: 253–266).

157 Hong, S. G., Y. Kim, and F. Venneri. 2008. Characterization of a Sodium-Cooled Fast Reactor in an MHR–SFR Synergy for TRU Transmutation. Annals of Nuclear Energy 35 (8): 1461–1470.

158 Ingersoll (2009: 589–603).

159 Kang, J.-H., N.-I. Tak, and M.-H. Kim. 2012. Thermo-Mechanical Analysis of the Prismatic Fuel Assembly of VHTR in Normal Operational Condition. Annals of Nuclear Energy 44: 76–86.

160 Karahan, A. 2010. Possible Design Improvements and a High Power Density Fuel Design for Integral Type Small Modular Pressurized Water Reactors. Nuclear Engineering and Design 240 (10): 2812–2819.

161 Kessides (2012: 185–208).

162 Lee, Y.-G., J.-W. Kim, and G.-C. Park. 2012. Development of a Thermal–Hydraulic System Code, TAPINS, for 10 MW Regional Energy Reactor. Nuclear Engineering and Design 249: 364–378.

163 Lim et al. (2011: 4079–4088).

164 Matsumura, T. et al. 2005. Core Performance of New Concept Passive-Safety Reactor "Kamado"—Safety, Burn-Up and Uranium Resource Problem. Progress in Nuclear Energy 47 (1–4): 131–138.

165 Şahin, S., and F. Sefidvash. 2008. The Fixed Bed Nuclear Reactor Concept. Energy Conversion and Management 49 (7): 1902–1909.

166 Sambuu and Obara (2012: 210–215).

167 Shropshire (2011: 299–307).

168 Slessarev (2007: 883–895).

169 Ibid., 636–646.

170 Tominaga et al. (2005: 106–114).

171 Toshinsky, G. I., O. G. Komlev, and K. G. Mel'nikov. 2011. Nuclear Power Technologies at the Stage of Sustainable Nuclear Power Development. Progress in Nuclear Energy 53 (7): 782–787.

172 Vujić et al. (2012: 288–295).

173 Wallace (2006: 445–453).

174 Yang et al. (2006: 2376–2385).

175 Yoon, H. J. et al. 2012. Potential Advantages of Coupling Supercritical CO_2 Brayton Cycle to Water Cooled Small and Medium Size Reactor. Nuclear Engineering and Design 245: 223–232.

176 Zhang and Sun (2007: 2265–2274).

177 Zheng, Y. J. et al. 2012. Thermal Hydraulic Analysis of a Pebble-Bed Modular High Temperature Gas-Cooled Reactor with ATTICA3D and THERMIX Codes. Nuclear Engineering and Design 246: 286–297.

178 Zrodnikov et al. (2008: 170–178).

179 Zrodnikov et al. (2011: 237–244).

3

SYMBOLIC CONVERGENCE

Hydrogen fuel cells and the engineering community

Hydrogen fuel cells have been promoted for many decades now on the grounds that they could enable a "hydrogen economy" where hydrogen, the most abundant element in the universe, is put to use to fuel cars and power plants and somewhat miraculously generate only water as "pollution." The International Association of Hydrogen Energy was established in 1973 to investigate the promise of hydrogen technologies and deployment pathways,[1] and in 1974 Germany, Japan, and the United States alone spent $10 billion (in unadjusted terms) on hydrogen research. In the years that followed, industrialized countries belonging to the OECD countries spent an additional whopping $20 billion from the 1970s to the 1990s[2,3] and the International Energy Agency estimated that the private sector contributed another $40 billion.[4]

Hydrogen, in other words, is serious business, or at least a credible arena of energy research. Even in the past few years, the promise of a hydrogen economy continues to attract significant attention among politicians, the media, and some academics,[5,6,7,8,9] although funding streams have diminished. In this chapter, I argue that an explanation lies in the way that the hydrogen economy fulfills psychological and cultural needs related to a future world where energy is abundant, cheap, and pollution free, a "fantasy" that manifests itself with the idea that society can continue to operate without limits imposed by population growth and the destruction of the environment.

The chapter, building on earlier research,[10] begins by explaining how a hydrogen economy could work, and it identifies a host of socio-technical challenges to explain why the creation of a hydrogen economy would present immense (and possibly intractable) obstacles. The chapter then depicts my research methodology consisting of two literature reviews and research interviews of energy experts (done during the most recent peak in hydrogen funding, about a decade ago). Next, the chapter elaborates on the application of symbolic convergence theory, a general communications theory about the construction of rhetorical fantasies. I employ symbolic convergence theory to identify five prevalent fantasy themes and rhetorical visions—independence, patriotism, progress, democratization, and inevitability—in academic and public discussions in favor of the hydrogen economy. I conclude by offering implications for scholarship relating to energy policy and science and technology studies more broadly.

By focusing on the hydrogen economy at the level of rhetoric and fantasy—and not merely hardware, people, research laboratories, and institutions—my study extends beyond individuals and technical artifacts to look intimately at discursive culture. Anthropologists and psychologists have long held that human behavior has many fundamentally irrational attributes.[11,12] Mass fantasies order ideas and values, and have the ability to shape behavior and developmental pathways. If it is true that fantasies relating to energy consumption and abundance have become publicly standardized community values, they will continue to mediate the experience of individuals until they are identified and critically examined. If taken to the extreme, values antithetical to consumption and abundance, such as conservation and environmental stewardship, could become socially unacceptable.

Conceptualizing and problematizing a "hydrogen economy"

Unlike conventional fuels, hydrogen is not an energy source but an energy carrier. Hydrogen is rarely found by itself the way other minerals (such as natural gas and petroleum) or resources (such as moving water, wind, or sunlight) exist. Instead, engineers and scientists have to "free" hydrogen from chemical compounds in which it is bound up. Two methods are employed to "liberate" hydrogen: using heat and catalysts to reform hydrocarbons or carbohydrates, or using electricity to split or electrolyze water. One of hydrogen's key strengths is that it can be an especially versatile energy carrier, stored as a gas or liquid, or attached to metal hydrates. Given its relative abundance, moreover, engineers and scientists can make it from almost any energy source and, conversely, convert it back to electricity.

The Institute for Nuclear Energy in Vienna introduced the concept of creating a "hydrogen economy" in 1971. The central premise was to create hydrogen using high temperature nuclear reactors that could then be used to replace fossil fuels. Proponents suggested that such an economy could utilize hydrogen fuel cells, small modular electrochemical devices similar to batteries, as a primary energy source for vehicles and power stations. As Figure 3.1 illustrates, hydrogen is undoubtedly alluring, given its versatility (and possible abundance) makes it a potential fuel source for multiple sectors (electricity, transport, industry) with multiple "value added applications" including vehicles, synthetic fuels, refining, and ammonia production.

However, a closer look suggests that significant socio-technical obstacles occur at virtually all stages of the proposed hydrogen economy fuel cycle, from production and distribution to storage and use. On the production side, using

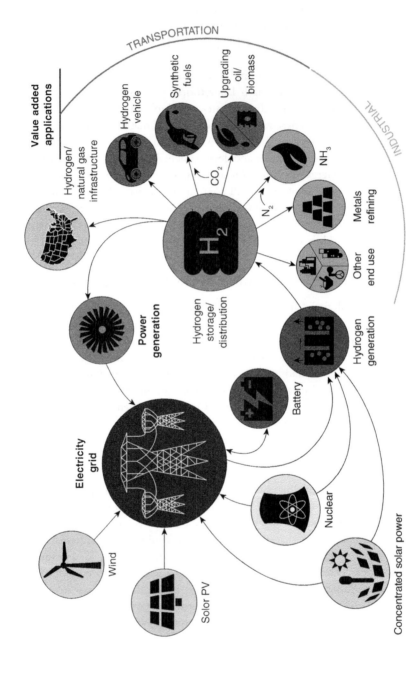

FIGURE 3.1 Visualizing the hydrogen economy in the United States

Source: U.S. Department of Energy

hydrogen as an energy carrier creates fundamental problems unavoidable by the laws of physics and thermodynamics. Hydrogen is a source of energy only if it can be taken in its pure form and reacted with another chemical, such as oxygen. Natural forces have already oxidized all the hydrogen on earth, with the exception of hydrocarbons, so that none of it is available as usable fuel. The rest has to be "made." Oil refineries use hydrogen to purify fuels, and chemical manufacturers employ it to make ammonia and other compounds. Both industries obtain a vast majority from high temperature processing of natural gas and petroleum. The method, however, sacrifices 30 percent of the energy content in the gas to obtain 70 percent of the energy in hydrogen, and emits prodigious amounts of carbon dioxide in the process—around 32 kilograms of carbon dioxide for one kilogram of hydrogen.[13] This small amount of hydrogen production already accounts for 2 percent of global energy demand.[14]

The second method, electrolysis, is equally problematic. To produce enough hydrogen through electrolysis to displace the oil needs of only the United States, 230,000 tons of hydrogen would need to be produced every day—enough to fill 13,000 Hindenburg blimps—and electricity generating capacity would have to double.[15] Similarly, the European Commission sponsored an intensive three-year study on the hydrogen economy and concluded that the use of hydrogen as a primary fuel would significantly contribute to climate change because of its greenhouse gas intensity.[16] Pathways for making hydrogen are exceptionally expensive, and the Commission concluded that energy prices would have to jump *ten times* existing rates before the hydrogen production techniques would be cost effective.

In terms of distribution and storage, long-distance hydrogen pipelines would need to be very large because gaseous hydrogen is so diffuse (less than one-third the energy content per unit volume as natural gas). Hydrogen pipelines would also rely on large amounts of energy to move the gas along the line. Such pipelines would have to be specially constructed, as hydrogen is irreconcilable with current infrastructure. Hydrogen storage is equally challenging. Storing hydrogen in its gaseous state requires large, high-pressure cylinders, demanding significant storage space. Although liquid hydrogen lessens the need for storage, scientists must super-cool it to temperatures below –423 degrees Fahrenheit. At these temperatures, about 40 percent of the energy in the hydrogen must then be expended to liquefy it. The American Physical Society went even further and concluded that a new material must be *invented* to solve the hydrogen economy storage problem.[17]

Both distribution and storage raise safety issues beyond their technical obstacles. Hydrogen is flammable at a much wider range of concentrations than natural gas, and hydrogen flames are barely visible.[18] For example, electrical storms have ignited hydrogen fuel from miles away, and dozens of people have died from unwittingly stepping into invisible hydrogen flames. One assessment of hydrogen use in the petrochemical industry found that undetected hydrogen leaks caused more than 20 percent of all hydrogen accidents, despite special training, protective clothing, and stringent safety regulations throughout the sector.[19]

Finally, challenges to the hydrogen economy exist at the retail and end-use level. General Motors estimates that providing national coverage for the first million hydrogen vehicles in the U.S. would require more than 12,000 hydrogen stations in cities and along interstates, each costing more than one million dollars.[20] If policymakers built the system to serve 100 million cars, its cost would exceed $1 trillion. Hydrogen vehicles, moreover, face difficulties relating to the operation of fuel cells, which currently degrade too quickly with use. The best available hydrogen fuel cell systems today cost $950,000 and last 2,000 hours whereas most commercial vehicles need a minimum lifetime of 5,000 hours and cost only $24,000.[21] Another survey of technologies available to reduce the risks of climate change ranked hydrogen second to last, noting that the world would need 30 times as much as today's current production of platinum just to make the proton exchange membranes.[22] Numerous other studies have concluded that hydrogen is neither the easiest nor the cheapest way to accomplish the goals of reducing air pollution or minimizing dependence on fossil fuels.[23,24,25,26,27,28] Even under a best-case scenario where researchers throw an unlimited amount of money into hydrogen research, commercialization would not likely occur until after 2035.[29]

Research design and empirical strategy

Because an investigation of the hydrogen economy, rhetoric, and visions involves technology, communication practices, people, institutions, and the naturalization and proliferation of mass fantasies, the chapter utilizes a unique methodology. Brent Brossmann and I began by documenting visions of the hydrogen economy as articulated by its proponents in the academic and popular literature during perhaps the height of its research popularity. Academic studies were first identified by searching various peer-reviewed journals using electronic journal databases with the terms "hydrogen" or "fuel cells" and "economy," "scenario," "roadmap," or "pathway," similar to the methodology employed by McDowall and Eames.[30] A total of 119 studies, published between 1999 and 2008, were reviewed, and were then narrowed for those that actually discussed the hydrogen economy (rather than merely using the words "hydrogen" or "economy") with a final sample of 34 articles. This time period of 1999–2008 was selected to reflect the peak in expectations (and funding) for hydrogen research, which indeed began to lose much of its glamour after 2009 and 2010.[31,32,33,34]

To assess the prevalence of hydrogen economy visions and narratives in the public sphere during this peak in funding, we searched the 50 most popular newspapers by subscription (such as the *New York Times* and *Wall Street Journal*), 50 most popular news magazines (such as *Newsweek* and *The Economist*), political testimony, and congressional transcripts from 2002–2007. A staggering 756 newspaper articles, 204 magazine articles, and 23 political transcripts were collected for a total of 983 documents. Because we did not particularly appreciate the prospect of having to read almost a thousand articles, we randomly selected one out of every 10, narrowing the field to 98 articles, 22 of which dealt with the hydrogen economy. Of these, a smaller sample of 18 stated a vision of the hydrogen economy. This literature review is depicted visually in Figure 3.2.

Table 3.1 illustrates some of the features of the final sample of 52 articles. Academic articles tended to come from the energy policy literature, notably journals such as *International Journal of Hydrogen Energy, Journal of Power Sources, Energy Policy, Energy*, and *Electricity Journal*. These articles were written predominately by authors affiliated with universities, energy companies, consulting firms, and government sponsored research laboratories. While their institutional affiliations represented more than 20 countries, half (17 out of 34) had authors or co-authors from the United States. Interestingly, however, a significant proportion of affiliations were also from countries in the developing world including Algeria, Brazil, Croatia, India, Kosovo, Nepal, Slovenia, South Africa, Turkey, and the Ukraine, implying that visions of the hydrogen economy are prevalent around the world, and also that the themes exhibit subtle but important differences based on their cultural context. My sample of the popular literature was mostly limited to the United States, with articles coming from political and media transcripts, magazines such as *Business Week* and the *New Scientist*, and newspapers such as the *Boston Globe, Houston Chronicle*, and *Atlanta Journal-Constitution*. This bias for sources in English does constrain the generalizability of the chapter's findings. The reliance on sampling also runs the risk of presenting somewhat anecdotal evidence instead of a deep and through depiction of the popular literature.

While the fantasy themes associated with hydrogen did display minor albeit important differences by country, they did not differ substantively by institutional affiliation or the academic training of the authors. One might expect that the

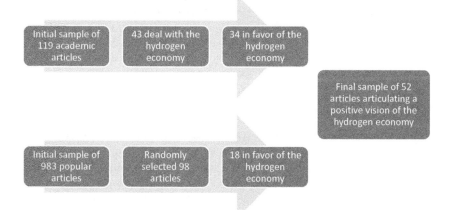

FIGURE 3.2 Literature selection process for academic and popular hydrogen articles
Source: Author, based on Sovacool, B. K., and B. Brossmann 2010, April[35]

TABLE 3.1 Details of my final sample of 52 articles in favor of a hydrogen economy

	Source	Institution/Publication
Academic articles	Ale and Shrestha (2008)[36]	Tribhuvan University (Nepal) and Western Michigan University (United States)
	Aranson and Sigfusson (2000)[37]	University of Iceland (Iceland)
	Balat (2008)[38]	Sila Science and Energy Company (Turkey)
	Baretto et al. (2003)[39]	IIASA (Austria) and Tokyo Electric Power Company (Japan)
	Blanchette (2008)[40]	Energy consultant (United States)
	Boudries and Dizene (2008)[41]	CDER (Algeria) and USTHB (Algeria)
	Brey et al. (2006)[42]	Hyergreen Technologies (Spain) and Pablo de Olavide University (Spain)
	Clark (2008)[43]	University of California (United States)
	Clark and Rifkin (2006)[44]	Green Hydrogen Scientific Advisory Committee (United States) and Foundation on Economic Trends (United States)
	Clark et al. (2005)[45]	Clark Communications (United States)
	Crabtree et al. (2004)[46]	Argonne National Laboratory (United States), Massachusetts Institute of Technology (United States), and Oak Ridge National Laboratory (United States)
	Doll and Wiet (2008)[47]	Fraunhofer Institute (Germany)
	Dunn (2001) [48]	Worldwatch Institute (United States)
	Dunn (2002)[49]	Worldwatch Institute (United States)
	Dutton and Page (2007)[50]	CCLRC Rutherford Appleton Laboratory (United Kingdom) and Institute for Transport Studies (United Kingdom)
	Edwards et al. (2008)[51]	University of Oxford (United Kingdom), Imperial College (United Kingdom), and Rutherford Appleton Laboratory (United Kingdom)
	Elam et al. (2003)[52]	National Renewable Energy Laboratory (United States), SunaTech (United States), Australian National University (Australia), Uppsala University (Sweden), and Norsk Hydro (Norway)
	Golstov and Veziroglu (2001)[53]	Donetsk State Technical University (Ukraine) and Clean Energy Research Institute (United States)
	Golstov et al. (2006)[54]	Donetsk State Technical University (Ukraine) and Clean Energy Research Institute (United States)
	Greyvenstein et al. (2008)[55]	Pebble Bed Modular Reactor Corporation (South Africa)

(Continued)

TABLE 3.1 (Cont.)

	Source	Institution/Publication
	Hodson et al. (2008)[56]	University of Salford (United Kingdom)
	Hotza and Costa (2008)[57]	Federal University of Santa Cataraina (Brazil) and University of Queensland (Australia)
	Lovins and Williams (1999)[58]	Rocky Mountain Institute (United States)
	Maack and Skulason (2006)[59]	University of Iceland (Iceland) and Icelandic New Energy (Iceland)
	Muradov and Veziroglu (2005)[60]	Florida Solar Energy Center (United States) and Clean Energy Research Institute (United States)
	Muradov and Veziroglu (2008)[61]	University of Central Florida (United States) and University of Miami (United States)
	Penner (2006)[62]	Center for Energy Research (United States)
	Saxena et al. (2008)[63]	Indian Institute of Petroleum (India)
	Sherif et al. (2005)[64]	University of Florida (United States) and University of Connecticut (United States)
	Taanman et al. (2008)[65]	Erasmus University Rotterdam (Netherlands), MERIT (Netherlands), and Eindhoven University of Technology (Netherlands)
	Talijan et al. (2008)[66]	University of Waterloo (Canada) and University of Ljubljana (Slovenia)
	Tseng et al. (2005)[67]	Department of Energy (United States) and Brookhaven National Laboratory (United States)
	Winter (2004)[68]	International Association for Hydrogen Energy (Germany)
	Zidansek et al. (2008)[69]	Jozef Stefan Institute (Slovenia), University of Ljubljana (Slovenia), University of Prishtina (Kosovo), and Ruder Boskovic Institute (Croatia)
Popular articles	Aston (2003)[70]	Business Week
	Abraham (2002)[71]	Department of Energy
	Berg (2008)[72]	Idaho Falls Post Register
	Boehlert (2003)[73]	House Science Committee
	Bush (2003)[74]	Office of the President
	Chandler (2005)[75]	New Scientist
	Clayton (2004)[76]	Christian Science Monitor
	Saillant (2002)[77]	Plug Power Incorporated
	Faulkner (2005)[78]	Department of Energy
	Ledford (2003)[79]	Atlanta Journal-Constitution
	Lovatt-Smith (2008)[80]	New Scientist
	Menchaca (2008)[81]	The Post and Courier

(*Continued*)

TABLE 3.1 (Cont.)

Source	Institution/Publication
Rifkin (2003)[82]	Foundation on Economic Trends
Rifkin (2005)[83]	Foundation on Economic Trends
Roper (2005)[84]	Houston Chronicle
Santa Fe New Mexican (2005)[85]	Anonymous
Schwartz and Randall (2003)[86]	Global Business Network and Shell Oil
Spowers (2003)[87]	New Statesman

Source: Author, based on Sovacool, B. K., and B. Brossmann 2010, April[88]

popular and media advocates would treat the hydrogen economy more favorably and less skeptically than the academic and technical advocates, but I found this not to be the case. As I document below, both groups subscribed to the five fantasy themes associated with the hydrogen economy. This type of uniformity may be explained by a cross pollination of ideas and references between the two groups, with the popular literature often referencing technical reports and books and the academic literature also citing newspapers and magazine articles. Moreover, the hydrogen economy seems to be a case where the lines between "public" and "academic" research are blurred, perhaps because the hydrogen economy is a technically complex issue and also because many of the academic advocates spend their time writing opinion pieces and popular articles. In this case, the line between technical and popular discussions is unclear.

Because my literature selection process was intended to give us a sample of articles in favor of the hydrogen economy, to get a fuller perspective I decided to supplement my review with research interviews of energy experts. 62 formal, semi-structured interviews were conducted at 45 institutions in the United States including electric utilities, regulatory agencies, interest groups, energy systems manufacturers, nonprofit organizations, energy consulting firms, universities, national laboratories, and state institutions. (Details about these interviews are presented in Appendix 3.1) Those interviewed possessed exceptionally diverse educational backgrounds in various fields, such as environmental engineering, business administration, economics, engineering and public policy, history, geography, law, chemistry, biology, environmental science, and energy resource management. Participants were asked both what they thought the most significant impediments were to technologies like hydrogen fuel cells, and whether they thought current government research and development would lead to a sustainable energy future. Interestingly, as will be documented below, these experts were overwhelmingly skeptical and critical of the hydrogen economy, although their comments may be slightly biased against hydrogen since many of the institutions are involved in promoting other technologies such as renewable energy and energy efficiency.

My own research interviews confirmed many of the problems discussed in the section above. When I asked these 62 professionals working in the energy policy field about which future energy technologies they believed held the most promise, not one of them mentioned hydrogen. Moreover, when asked whether current research funds expended on hydrogen would lead to a sustainable future with technologies providing cheap and abundant energy, 94 percent said "No." Three participants even went so far as to refer to such research as "atrocious public policy," "nibbling only at the margins," and a "dismal excuse for comprehensive energy solutions."

Despite all of these substantial challenges, however, hydrogen remained at the forefront of international energy policy and research and development in the 2000s. Figure 3.3 shows that research expenditures on hydrogen rose significantly from 2002–2007. In the next section, I will explore the concept of "fantasy themes" and "symbolic convergence theory" to help explain why.

Fantasy themes and symbolic convergence theory

Symbolic convergence theory has its roots in psychological studies of Sigmund Freud, Jacques Lacan, and Robert F. Bales. Clinical work undertaken by psychoanalysts such as Sigmund Freud and Jacques Lacan relate to idea that fantasies

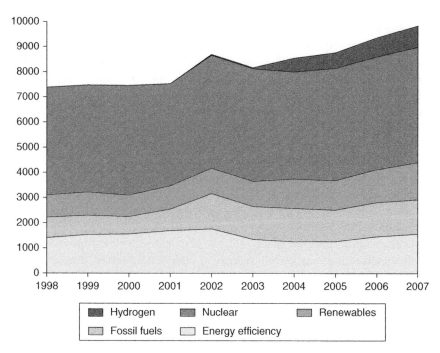

FIGURE 3.3 Research and development expenditures among the members of the International Energy Agency (in millions of U.S. Dollars), 1998 to 2007

Source: International Energy Agency

serve important social functions by allowing the mind to interpret and deal with the complexities and stresses of modern life.[89,90] The problem, however, is that fantasies themselves are often hidden from those that have them, buried deep within the subconscious. People seldom act upon their fantasies, leading to frustration and transference, a process whereby people locate fantasies not just in people and actions but also objects. In addition, large numbers of people can share group fantasies. The later works of both Freud and Lacan focused on finding the emergent fantasy "themes" that were shared across different social groups.[91,92] (Any reader that has ever fantasized about winning the lottery, harming their boss, finding a perfect soulmate, racing in the Indianapolis 500, being presented with a Nobel Prize, or being elected President or Prime Minister may already know about this tendency for similar fantasies to become shared or to resonate across diverse audiences.)

Most significant here is the work of Robert F. Bales, who conducted numerous studies on small group interaction and found that in every group dynamic the sharing of fantasies occurred.[93] Bales concluded that group fantasies correlated with individual fantasizing, emphasizing the irrational and unconscious elements of sharing fantasies, and that groups often make collective decisions that they never would have agreed to on an individual basis once such fantasies have "chained" out (diffused to a larger group). Indeed, Bales even found that group members had a tendency to become emotionally committed to a certain fantasy—ultimately changing their behavior in attempts to fulfill it.

Symbolic convergence theory builds on these earlier ideas to create a general theory of communication. First introduced by Ernest G. Bormann in 1972 and further developed by, and with, his colleagues John F. Cragan and Donald D. Shields, symbolic convergence theory looks at the collective sharing of fantasies and how group consciousness affects human action. [94,95,96,97,98,99,100] The theory does this by demonstrating the communicative influence of fantasy, a force that continuously affects the consciousness of individuals, groups, and large publics. Like "gravity," symbolic convergence theory posits that fantasy is a critical element of human interaction, a constant needed in order for humans to explain and interpret their experiences.[101] In short, the theory aims to account for the processes that "create, raise, and sustain a rhetorical community's consciousness."[102]

In this context, the term fantasy is not to be confused with the common usage implying something completely imaginary or distinct from reality. Instead, fantasy is a way that communities of people share their social reality. In this sense, the term employs a meaning much closer to the creative interpretation of events that fulfill a psychological and rhetorical need.[103] Fantasies may create symbolic realities that encompass ideologies, orientations, visions, and worldviews, but while these are immaterial they are still "real" to many people. Symbolic convergence theory has many interrelated concepts and terms, but the four most important ones for this chapter are *fantasy themes and types, dramatis personae, symbolic cues,* and *rhetorical visions.*

Fantasy themes and types tap into common experience or interpretations and have a capacity to prompt shared interpretations of those experiences among groups of people.[104,105] In other words, fantasy themes enable different dramatic events to be interpreted similarly. These often have two important elements: *dramatis personae* and *symbolic cues*.

Fantasy themes often take the form of *dramatis personae*, which is to say that they have specific dramatic elements such as plot lines, stories, and characters. The theory posits that human beings use drama and fantasy to interpret everyday actions and experiences at the ontological level, implying that fantasy themes are inseparable from human thought. Human beings seldom understand events in all their complexity. Instead, they try to understand things in relation to familiar people and storylines. At heart, the theory holds that humans are storytellers, and that when they share a dramatization of an event, they make sense out of its complexity by creating a script to account for what happened. Fantasy fulfills a basic psychological need to understand experience in terms of human desire, ambition, and action.[106]

People become enrolled in fantasies the same way they become attached to particular jokes, engrossed in a novel, identify with a particular movie, or believe a sacred myth. Such dramatizing enables groups of people to come to a "symbolic convergence" about that part of their common experience.[107] If the message is humorous, the audiences laugh; if the speaker presents a leading character in a sympathetic way, the audience sympathizes. Dramas can thus not only chain out within small groups or individuals but also can be created and sustained by the technologies of mass media (such as academic journals or newspaper articles) to large groups that then distribute such fantasies back to smaller groups and individuals, creating a shared reality.[108]

Frequently, as part of this dramatic storyline, members subscribing to a particular fantasy theme will develop or reuse code words, phrases, slogans, or nonverbal signs and gestures. These *symbolic cues* trigger previously shared fantasies.[109] The cues may refer to a geographical or imaginary place or the name of a persona, and they may arouse tears or evoke anger, hatred, love, affection, laughter, and a range of other emotions. When a community of people comes to share a fantasy type, they will frequently respond to general statements alluding to the type without the supporting references to specific themes.

When groups share certain fantasy themes, along with their dramatis personae and symbolic cues, the individual themes often coalesce into a larger narrative or discourse known as a *rhetorical vision*. That vision represents the consciousness of the fantasy theme's adherents, creating a "rhetorical community" with its own distinct worldview.[110] Rhetorical visions often adhere to one of three master analogues. They can be righteous, describing a correct or moral way of doing things. Rhetorical visions of this nature reflect an interest in right or wrong, appropriate or inappropriate behavior, just and unjust action. They can be social, emphasizing important human relationships such as trust, loyalty, and brotherhood. They can be pragmatic, underscoring efficiency and cost-effectiveness. And, at times, they can weave together all three analogues.

Bormann, for example, found a "fetching good out of evil" fantasy common to communities that attempted to look for the good in times of great disaster or war.[111] Scharf found a rhetorical vision of psychiatrists as scapegoats, rather than lawyers or criminals, for public indignation with the guilty by insanity verdict.[112] Looking at slightly larger groups, Bormann argued that Democratic presidential candidates always portrayed Republicans as consorting with wealthy multinational corporations to formulate policies that hurt the poor.[113] In contrast, Republican candidates portrayed Democrats as fiscally irresponsible for throwing money at social problems.

Rhetorical visions can become so encompassing and impelling that they permeate a group's entire social reality in almost all aspects of living. During the Cold War, Bormann, Cragan, and Shields argue that political leaders at all levels of American government combined four fantasy themes into a coherent rhetorical vision depicting the evilness of communism and the former Soviet Union[114]: (a) the idea that totalitarian forces conspired to conquer, dominate, and enslave the world; (b) the belief that aggressor communist nations could be appeased; (c) the notion that battles of the upholders of freedom and dignity against the forces of enslavement and domination are just wars; and (d) the ideal that after the great war lasting peace based on world law would emerge.

A similar rhetorical community existed in Germany before World War II. Bormann argues that Adolf Hitler was such a strikingly successful speaker because he gave voice to what many Germans already believed to be true.[115] Hitler successfully created a rhetorical vision consisting of three separate fantasy themes: (a) that Germany had a great destiny; (b) its economic problems were not the fault of its loyal citizens; and (c) the country, acting together, could do something about it. Hitler's rhetorical vision portrayed and represented the German population more than it "conquered" them. Bormann notes that from a purely technical standpoint, many of Hitler's speeches forming this rhetorical vision were chaotic, contradictory, and senseless when taken literally. Hitler's rhetorical vision, in contrast, housed deeper meaning that began in tones of deep pessimism but ended with overjoyed redemption, touching many Germans in ways that purely rational and logical speech could not.

Symbolic convergence theory offers at least three theoretical benefits. First, it reveals the link between small group action and mass-mediated communication processes that are critical in motivating people to act on common values. Fantasy themes may start with a few individuals and chain out to small groups, but once they enter the public sphere they are transmitted to thousands of other individuals and groups that internalize them and start the process again. Thus, the theory depicts how mass consciousness can be attained.[116] Instead of moving simply from media to opinion and then to the public, the theory posits that information flows in all directions between all agents creating a web of interaction and making possible a unified rhetorical vision.[117]

Second, the theory focuses intimately on the symbolic aspects of communication, highlighting that reality can be created through symbols, metaphors, and images. The theory is thus well attuned to both logical and emotional qualities of rhetoric and the aesthetic and technical aspects of communication. The theory emphasizes that language is both rational and allegorical, reflecting the ability to story-tell and construct myths as much as to reason.

Third, the theory focuses simultaneously on the content of the message and the audience. Even though meaning, emotion, and value all lie in the message, the message is constituted with the audience, suggesting that the locus of meaning is jointly created by both.[118] This view sharply contrasts with the Aristotelian view that the meaning existed in the speaker that picked the words to create a message, or McLuhan's view that the medium conveying the message provides meaning. Rather, symbolic convergence theory incorporates all of these aspects: communication is about the message, who says it, its content, its form, and the audience.

Perhaps because of these benefits, scholars have used symbolic convergence theory to look at fantasy themes and mass fantasies on a wide range of subjects in a rich variety of disciplines. Ronald Bishop found three fantasy themes surrounding the popular children's television show *Mr. Rogers' Neighborhood.*[119] James W. Chesebro used symbolic convergence theory to analyze how homosexuals are regarded among social scientists.[120] John F. Cragan and Donald C. Shields utilized fantasy theme analysis to reveal how most foreign policy experts viewed Russia as a villain on the international stage during the Cold War.[121] Kristina Drumheller employed the theory to investigate how teenagers express religious dogma.[122] Margaret Duffy used it so show how hate groups perceived of minorities in online chat rooms.[123] Economists have relied on it in market-based research to identify and test the efficacy of fantasy themes in rural communities used to attract medical physicians.[124] The U.S. Department of Education has wielded symbolic convergence theory to assess the persuasive force of the "Just Say No" anti-drug program.[125] Fortune 500 companies have deployed it to strategically reposition themselves by using the theory's principles to segment markets and test the power of advertising and sales messages.[126,127] And nonprofit groups and educators have used it to identify messages that students would retain, expand enrollment, and improve parental involvement,[128] as well as to enhance teachers' bargaining power when negotiating salary increases, labor protests, or strikes.[129]

The fantasy themes and rhetorical visions of a hydrogen economy

When applied to discussions about the hydrogen economy, symbolic convergence theory reveals a rich number of fantasy themes and visions, each with their own different *dramatis personae*, along with recurring *symbolic cues* and a collective *rhetorical vision*.

More specifically, I found (1) a theme of *inevitable destiny* that depicts hydrogen as the inescapable and unavoidable result of socio-technical development; (2) a theme of *energy independence* where advocates see hydrogen technologies as offering countries a robust, domestically insulated energy infrastructure immune from the vagaries of the global energy marketplace; (3) a theme of *patriotism* that paints hydrogen as a way to achieve national leadership, competitiveness, strength, and vitality; (4) a theme of unlimited *progressive growth* that views hydrogen as a mechanism to achieve endless economic expansion fueled by pollution-free and limitless supplies of energy; (5) and a theme of *energy democratization* that sees hydrogen as ushering in a wave of decentralized energy production and use.

Inevitable destiny

The first fantasy theme articulated by supporters is the belief that the hydrogen economy is inevitable. The theme is touted repeatedly in the professional literature, whether it is from a historical perspective,[130] because shortages in the supply of conventional fossil fuels will force it,[131] or because hydrogen "is the indispensable Kyoto compatible, clean and abundant energy carrier."[132] Muradov and Veziroglu argue that "the dominant role of hydrogen in a sustainable energy future is widely accepted."[133] Blanchette assumes its dominance by claiming that "hydrogen … will come to the fore within the first half of the twenty-first century."[134]

The argumentative function of this belief cannot be overstated, for if the hydrogen economy is inevitable, then the only obstacles are transitory, opposition is ultimately fruitless, and the logical conclusion is to expedite the commitment to hydrogen so that the benefits may be quickly maximized. The inevitability argument, however, does not deny that the transition requires a fundamental shift in paradigms. This paradigmatic transformation involves two parts, the movement away from fossil fuels and the movement toward the hydrogen economy.

The first step, the search for alternatives to fossil fuels, has clearly occurred, but the fantasy theme assumes that the second element is also underway, a search "to find new clean sources of fuels to convert to hydrogen in both economical and efficient ways."[135] Although the transition to hydrogen is a "paradigm change of enormous magnitude" which accentuates "the need for a robust and well thought out transition,"[136] the process is simplified by the belief that the general details of the plan have existed for more than twenty years.[137] According to this theme, given that the hydrogen economy will occur eventually, the only question is how to make it occur sooner rather than later. The theme calls upon consumers and users to adopt hydrogen technologies in a rational and unprejudiced manner, and completely erases the technical challenges inherent with transitioning to a hydrogen economy.

Energy independence

By far the most common theme depicted was that of energy independence and enhanced energy security. Clean, hydrogen powered automobiles and power stations are seen as a way to minimize and even eliminate costly dependence on foreign sources of fuel, all the while contributing to economic growth. Berg discusses the role hydrogen can play in reducing dependence on foreign oil,[138] and Blanchette argues that improving security of energy supply is the real force behind the move to hydrogen.[139] Greyvenstein et al.,[140] Muradov and Veziroglu,[141] and Balat[142] also expand upon this theme to connect hydrogen use with improved national security.

One visible source for this theme is high-ranking political officials. United States President George W. Bush stated in 2003 that "one of the greatest results of using hydrogen power, of course, will be energy independence for this nation."[143] The fantasy theme therefore offers currently fuel-dependent countries a pathway to economic revitalization. Spencer Abraham, then Secretary of Energy in the United States, said as much in 2002 when he stated that:

> Imagine a world running on hydrogen later in this century: Environmental pollution will no longer be a concern. Every nation will have all the energy it needs available within its borders. Personal transportation will be cheaper to operate and easier to maintain. Economic, financial, and intellectual resources devoted today to acquiring adequate energy resources and to handling environmental issues will be turned to other productive tasks for the benefit of the people. Life will get better.[144]

By minimizing and eventually displacing imports of foreign fuels, the theme paints the hydrogen economy as an important tool to reduce national deficits, insulate economies for fuel shocks, and improve economic vitality. At the extreme, Sherif et al. note that hydrogen and electricity "can satisfy all the needs of human kind and form an energy system that would be permanent and independent of energy sources."[145]

Patriotism

This theme, hydrogen as patriotism, applies almost exclusively to the United States, although this could be because of the prominence of U.S. authors in my sample. In my review of the popular literature, many of the articles referenced an Apollo style commitment to developing the hydrogen economy as a way to renew American technological leadership and strength.[146,147,148,149] This part of the fantasy is particularly powerful since it undergirds a variety of fantasy chains. The comparison immediately relegates technological issues to the background since Kennedy's Apollo commitment demonstrated the ability to accomplish the "technologically impossible" more than 50 years ago. As Schwartz and Randall argue, "adopting Kennedy's 10-year time frame may sound absurdly optimistic, but it's exactly the kick in the pants needed to jolt the US out of its crippling complacency when it comes to energy. ... The good news is that the technical challenges are issues of engineering rather than science. That means money can solve them."[150]

However, such a fantasy also justifies a second chain that demands a public-private partnership behind a political commitment to creating the hydrogen economy. As Rifkin explains, "A hydrogen energy regime requires a public-private partnership on a grand scale, the kind we marshaled in the Apollo space program when the United States set its sights on landing a man on the moon in 10 years."[151] This combination of public and private financing is a well-developed line of argument within both the popular and professional literature. There is clear tension between the fantasy's claim that the adoption of a hydrogen economy is inevitable and the necessity of a well-funded public-private partnership to generate both a commitment and trillions of dollars to bring the hydrogen economy into existence. The answer is found in a timing argument that "without aggressive energy and environmental policies, the hydrogen economy is likely to emerge along the more incremental path, and at a pace that is inadequate for dealing with the range of challenges posed by the incumbent energy system."[152] The result is a fantasy theme encompassing the best of both worlds—the inevitability to reject counter arguments and the necessary obligation to justify calls for a major commitment from public and private actors.

Both chains expand into a call for action grounded in patriotism. The American space race was as much an expression of national pride as a technical challenge, and that is not lost on those who create the fantasy. The effort to achieve "energy independence through hydrogen is a patriotic duty"[153] and Bush's willingness to "commit the resources to help develop the technology for a hydrogen economy" clearly "deserves applause from across the political and ideological spectrum."[154] The visionary comparison to the success of the moon race seamlessly combines a pressing technological challenge, the need for public-private partnership, and a patriotic call to duty.

Some of the more recent articles from my sample have begun to expand this theme to other countries. Hodson observes that "the transition to a hydrogen based future is about securing national competitiveness" for not only the United States but also Canada and Iceland, and that each of these countries are concerned about "being a first mover" and not "falling behind in the competitive race."[155] Greyvenstein et al. comment that hydrogen can improve South Africa's role in the global economy,[156] Hotza and Costa that hydrogen could revitalize the economy in Brazil,[157] Edwards et al. that it can bring the United Kingdom into a new economic and political era.[158]

Progressive growth

This theme envisions hydrogen as instrumental to maintaining current levels of consumption and economic growth in the industrialized and industrializing world. It notes that although industrial growth has so far been powered by fossil fuels, the fact that these fuels are non-renewable means that a conventional model of energy production and use has always implied its own destruction, even when conveniently ignored. The theme highlights the environmental harms of burning fossil fuels (global warming, pollution), the impracticality of voluntarily reducing consumption (conservation), and the inevitability of resource limitations. One of the key rhetorical fantasies created by this theme is the ability to break that relationship, to provide a fuel that promises unlimited growth and no environmental degradation.

This fantasy theme of abundance and limitless consumption centers on the claim that hydrogen is the "forever fuel" because it never runs out, and when used to produce power the only byproducts "are pure water and heat."[159] Given the international concern over global warming, removing carbon from our energy supply is a tremendously powerful part of the fantasy. We are told that the combination of CO_2 and warming are "one of the main drivers for the attainment of [a] solar-hydrogen economy,"[160] that it "automatically solves, in principle, the global problem of the greenhouse effect,"[161] that it is "the ultimate step in climate stabilization,"[162] that it will "contribute to the reduction of energy-linked environmental impacts, including global warming,"[163] and that it "could be crucial to the future of mankind and the planet it ever-more-tenuously occupies."[164] Menchaca asserts that "hydrogen-powered vehicles generate zero emissions because the only byproduct is water vapor so pure you can drink it"[165] and Balat reiterates that "the only product of hydrogen combustion is water, which can be used in a circular way so as to realize a benign cycle of nature."[166] Edwards et al. claim that hydrogen is the key to "a green revolution in transportation,"[167] and Muradov concludes that "there exists a viable green or carbon-neutral path from current fossil-based to future hydrogen economy without disturbing a fragile environmental balance."[168]

In fact, the fantasy proclaims that the completion of the transition to the Hydrogen Civilization "will not only solve the world's environmental problems, but will also produce changes in the biosphere and its functioning over the historical and geological scales of time."[169] Such a vision not only provides warrants for our support of hydrogen energy, but implies that opposition to the vision invites ecological disaster. Balat even goes so far as to claim that "hydrogen holds the promise as a dream fuel."[170]

Not all of the literature is this hyperbolic. The environmental benefits of hydrogen are entirely dependent upon using renewable or other sources of clean energy in the conversion processes. However, arguments as to how a hydrogen economy could increase warming and pollution are rare. The vision is challenged by details such as the inefficiency of using natural gas as a transitional fuel as we search for ways to extract pure hydrogen from water, or in the warnings that we should not base policy on unproven technologies, such as the desire to unlock the "mysteries that Nature has long kept hidden" so that we can replicate the process "that allows plants to split water at room temperature using sunlight."[171] But it is here that the earlier rhetorical visions return with force. The details are not reasons to reject the vision, for it has already been established that the barriers are not insurmountable, that the reward is nothing less than the salvation of the planet, and that all we need to overcome these trivial concerns is leadership, money and commitment. From this perspective, legitimate concern about the viability of the hydrogen energy economy become little more than new justifications for further funding.

Energy democratization

A final fantasy theme describes the hydrogen economy as a path towards community empowerment and democratization by engendering a decentralized and more localized world. This theme envisions hydrogen as promoting more pluralistic, participatory, and community-owned forms of energy production. Some proponents even go so far as to frame the hydrogen economy as a fundamental altering of ecological values, changing the way that humanity conceptualizes its relationship with energy technologies and the environment.[172] A more moderate version of this theme shifts from arguing that hydrogen empowers people to a more nuanced debate over the merits of decentralized energy supply.

The fantasy theme sometimes begins by drawing parallels to the ascendancy of the World Wide Web. To proponents of this vision, the key to replacing the current energy infrastructure is decentralization, or more particularly the advent of the "energy internet" or the "Worldwide Energy Web." The goal is to combine the power of a decentralized electrical power grid with "hypercars," ultra-light vehicles running on hydrogen fuel cells. The vision reverses the traditional notion that our transportation fleet consumes energy. By combining a vision of decentralized energy generation with the preservation of the power grid, proponents depict a world in which cars consume energy for less than 5 percent of their existence (the time actually spent traveling) and generate electricity for more than 95 percent of their existence, as they are plugged directly into the power grid when otherwise not in use. The result is cars will power buildings, buildings will power cars, and the "larger commercial and industrial complexes" will trade their power surpluses in the spot market.[173]

Taken to its logical extreme, this decentralization of energy production transforms society. It will "make possible a vast redistribution of power" eliminating the "centralized, top-down flow of energy, controlled by global oil companies and utilities."[174] The result is peer-to-peer energy sharing, analogous to file-sharing on the internet, forcing energy companies to cooperate "or follow the evolutionary path of the dinosaurs."[175] Interestingly, here the fantasy has fallen victim to entelechy, the overdevelopment of a line of thinking, for taken to this extreme, the complete decentralization of energy production could harm the very businesses (electric utilities, grid operators, energy companies) that would need to catalyze investment in it.

Symbolic cues

In addition to fantasy themes, I found three recurring narratives or statements that can be regarded as symbolic cues, used by supporters to trigger fantasy themes about hydrogen. (This is not to say these claims are false, only that they represent important cues). Hydrogen is continually cited as the most abundant element in the known universe; it is often hailed as odorless and colorless; and it is frequently described as pure, being the simplest element. Upon closer inspection, however, each of these symbolic cues is less meaningful than they may appear.[176] Hydrogen may be abundant in the universe, but this fact has questionable relevance for use on Earth, where it is far from the most abundant element. The colorless and odorless aspects of hydrogen are probably a negative attribute for a widespread fuel since leaks will be difficult to detect. The simplicity of hydrogen as an element also has little relevance to its use as an energy resource.

Rhetorical visions

As is obvious by now, each of the five fantasy themes concerning the hydrogen economy—inevitability, independence, patriotism, progress, and democratization— have many different elements and, at times, contradict. As one example, the theme of *energy democratization* sees hydrogen as fundamentally altering the energy system and human relations with it, while the theme of *progress* sees it doing the opposite and enabling society to continue on its consumptive course. Furthermore, many of the articles in my sample promoted a variety of different themes all at once.

While each of the five fantasy themes was prevalent in both academic and popular discussions of the hydrogen economy, the specific nature of the themes did change based on the national affiliation of the authors. Consider the theme of independence. Because Nepal is rich in hydroelectric resources, the hydrogen economy is seen as a way to create Nepalese energy independence through large-scale electrolysis using hydroelectric reservoirs built along perennial rivers.[177] Because Algeria has a large surplus of fossil fuels and an abundance of sunshine, their hydrogen economy would create independence by initially tapping oil and gas reserves before transitioning to the use of non-tracking solar photovoltaic arrays.[178] In South Africa, which has plentiful coal reserves and a robust nuclear power sector, energy independence would be achieved through hydrogen production from coal liquids and modular pebble bed nuclear reactors, although this would conflict with other dimensions of the fantasy relating to environmental progress and democratization.[179] In Brazil, the hydrogen economy would be powered by biofuels such as ethanol and renewable resources such as wind and hydropower.[180] In India, the hydrogen economy would come from biomass and biogas.[181] In the Netherlands, it would come from natural gas.[182] Indeed, these subtle alterations on the theme of independence imply that the overall vision of a hydrogen economy varies based on local context where different motivations and expectations exist.

Despite these differences, however, the overall *rhetorical vision* of the hydrogen economy displayed some remarkable commonalities. Hydrogen is perceived as abundant and cheap by each of the themes, and also as a mechanism to replace all conventional forms of energy production. The vision is also incredibly vague, possessing immense strategic flexibility concerning particular hydrogen technologies and configurations. Moreover, the rhetorical vision of the hydrogen economy weaves together all three master analogues: it is righteous, emphasizing at times the patriotic or moral duty to invest in hydrogen; it is social, emphasizing how human relationships between each other and the natural environment can be reshaped by a hydrogen economy; and it is pragmatic, underscoring the rational efficiency and cost-effectiveness of a hydrogen transition.

Interestingly, each of the fantasy themes depicted here have different villains. The villains in the theme of *inevitability* are those, often through ignorance and self-centeredness, which doubt the legitimacy of the hydrogen transition. The villains in the theme of *independence* are oil suppliers and cartels such as OPEC that desire to raise energy prices. The villains in the theme of *patriotism* are those that seek to waste energy and select inefficient technologies. The villains in the theme of *progress*, interestingly, are split. Environmentalists see big industrial emitters of greenhouse gases as culprits threatening the vitality of the climate, whereas industrial emitters see environmentalists as villains seeking to constrain growth and place limits on industrialization. The theme of progress is able to give each group what they ostensibly want: a world of pollution-free and therefore limitless energy use. The villains in the theme of *democratization* are the integrated energy companies and electric utilities seeking to retain their control over energy production.

Conclusions

Notwithstanding all of the socio-technical challenges faced by a transition to the hydrogen economy—that hydrogen must be manufactured to become an energy carrier, that it is hazardous and unsafe, that current pathways of hydrogen production are expensive and energy-intensive, that the pool of experts I interviewed were extremely skeptical, and that widespread commercial use of hydrogen would require trillions of dollars of investment and technological breakthroughs that have yet to occur—a significant number of politicians, entrepreneurs, members of the public, and even academics continue to be captivated by its allure. My own review found 52 academic and popular articles envisioning a hydrogen future written by authors in more than 20 countries, including visions articulated by senior heads of state such as former President George W. Bush and Secretary of Energy Spencer Abraham. More than $70 billion in public and private research funds have been spent on hydrogen research from 1973–2007, and annual research expenditures on hydrogen were significant during this time among the members of the International Energy Agency.

What, then, explains the attraction? I believe that symbolic convergence theory, a framework describing how fantasy themes coalesce into rhetorical visions that can become widely shared, offers an answer. The challenges faced by a hydrogen transition are discounted in the face of the much more powerful and compelling fantasies associated with the hydrogen economy. Advocates dismiss attacks on the hydrogen economy as "unimaginative" and "premature" and instead subscribe to a grander vision permeated by fantasy themes surrounding conceptions of an inescapable hydrogen future, robustly independent energy sectors, revitalized national strength, accelerated technological and material progress, and decentralized energy supply. To be sure, the fantasy themes of inevitability, independence, patriotism, progress, and democracy have penetrated intellectual and public discourse about the hydrogen economy.

To say that hydrogen economy fantasies circulate, however, does not mean that they are universally shared or embraced, that those fantasies remain homogenous or pure among those that share it, or that those enthralled by such rhetorical visions are delusional. My own sample of the technical literature found 34 of 43 articles in favor of the hydrogen economy, whereas a strong majority of the 62 energy experts I interviewed were not swept up in the same rhetorical vision as the hydrogen advocates. One explanation is that scholars tend to write in favor of their own projects but not to position themselves against others. The problem with that practice is that it creates an illusion of unified thought when in fact none may exist. Even within hydrogen advocates, some of these actors see near-term and niche applications for hydrogen rather than the potential for radical and transformative change.[183,184] Some of the particular fantasy themes, such as energy independence, differ notably within national contexts and likely further across (and within) professional groups.

Nonetheless, and lastly, the chapter has shown that the fantasy themes surrounding the hydrogen economy have proliferated to a surprising number of people, including those in the developing world, and also a broad base of advocates from universities, government bodies, research laboratories, and the media. Such themes are not located in any one individual or group, nor are they confined to a particular type of hydrogen technology, and instead manifest themselves as a mass fantasy shared by stakeholders indifferent across countries and cultures. This implies that the provocative force of the hydrogen economy fantasy can, in the right circumstances, transcend any specific location or institution. While certainly not uniform, each of the underlying fantasy themes surrounding hydrogen serve cultural, psychological, and economic needs, whether they are immunity from dependence on foreign sources of fuel and macroeconomic dislocation (independence), pride and national identity (patriotism), continued economic expansion (progress), a need for change (democratization), or uncertainty concerning the future (inevitability). The demand for these sorts of fantasies will likely continue even if the hydrogen economy does not come to fruition.

APPENDIX 3.1 Research interviews

Name	Title	Institution	Date
Adam Serchuk	Senior Renewable Energy Program Manager	Energy Trust of Oregon	21 February 2006
Alan Crane	Senior Program Officer	National Academies of Science	2 February 2006
Alex Farrell	Assistant Professor, Energy Resources Group	University of California Berkeley	26 October 2005
Anonymous Expert	High-ranking Executive	Large, American Independent Power Producer	4 March 2006
Art Rosenfeld	Commissioner	California Energy Commission	6 December 2005
Benjamin Cohen	Assistant Professor, Science and Technology Studies	University of Virginia	6 March 2006
Brian Castelli	Executive Vice President	Alliance to Save Energy	7 February 2006
Brice Freeman	Project Manager, Distributed Generation	Electric Power Research Institute	22 February 2006
Chris Namovicz	Operations Research Analyst	U.S. Energy Information Administration	9 February 2006
Christopher Russell	Director, Industry Sector	Alliance to Save Energy	9 February 2006
Chuck Goldman	Group Leader, Electricity Markets and Policy Group	Lawrence Berkeley National Laboratory	25 October 2005
Claudia J. Banner	Senior Engineer, Renewable Energy Planning	American Electric Power	6 March 2006
Dave Stinton	Combined Heat and Power Program Manager	Oak Ridge National Laboratory	2 August 2005
David Baylon	President	Ecotope Consulting, Research, and Design	15 August 2005
David Garman	Undersecretary of Energy, Former Assistant Secretary for Energy Efficiency and Renewable Energy	U.S. Department of Energy	22 February 2006

(*Continued*)

Name	Title	Institution	Date
David Hawkins	Lead Industry Relations Representative	California Independent System Operator (ISO)	13 October 2005
David Hill	Deputy Laboratory Director, Science and Technology	Idaho National Laboratory	3 August 2005
Edward Vine	Staff Scientist	Lawrence Berkeley National Laboratory	25 October 2005
Irene Leech	Professor, Department of Resource Management	Virginia Polytechnic Institute and State University	27 March 2006
Jack Barkenbus	Executive Director, Energy, Environment, and Resources Center	University of Tennessee	4 August 2005
James Gallagher	Director, Office of Electricity and Environment	New York State Department of Public Service	4 October 2005
Jan Harris	Project Manager	Vermont Energy Investment Corporation	29 September 2005
Janet Sawin	Director, Energy and Climate Change Program	Worldwatch Institute	7 April 2006
Jeff Jones	Operations Research Analyst	U.S. Energy Information Administration	8 February 2006
Joe Catina	Manager of North American Projects	Ingersoll Rand Company	20 September 2005
Joe Loper	Vice President for Research and Analysis	Alliance to Save Energy	7 February 2006
John Plunkett	Partner and Former President	Optimal Energy Consulting	29 September 2005
John Warren	Director, Division of Energy	Virginia Department of Mines, Minerals, and Energy	16 September 2005
John Wilson	Senior Researcher	California Energy Commission	6 December 2005
Judi Greenwald	Director of Innovative Solutions	Pew Center on Global Climate Change	13 February 2006
Ken Tohinaka	Senior Energy Analyst	Vermont Energy Investment Corporation	29 September 2005
Larry Papay	Former Senior Vice President	Southern California Edison	21 September 2005

(*Continued*)

Name	Title	Institution	Date
Marilyn Brown	Commissioner	National Commission on Energy Policy	4 August 2005
Mark Hall	Senior Vice President	Primary Energy	5 October 2005
Mark Levine	Division Director, Environmental Energy Technologies	Lawrence Berkeley National Laboratory	25 October 2005
Michael Pomorski	Senior Associate	Cambridge Energy Research Associates	21 January 2006
Michael Karmis	Director	Virginia Center for Coal and Energy Research	16 February 2006
Paul DeCotis	Director of Energy Analysis	New York State Energy Research and Development Authority	4 October 2005
Paul Gilman	Former Assistant Administrator for Research and Development	U.S. Environmental Protection Agency	3 August 2005
Ralph Badinelli	Professor of Business Information Technology	Virginia Polytechnic Institute and State University	6 March 2006
Ralph Loomis	Executive Vice President	Exelon Corporation	6 October 2005
Rodney Sobin	Innovative Technology Manager	Virginia Department of Environmental Quality	16 September 2005
Ryan Wiser	Staff Specialist, Renewable Energy Systems	Lawrence Berkeley National Laboratory	25 October 2005
Sam Fleming	Former Program Manager, Renewable Energy Systems	Nextant Incorporated	27 October 2005
Scott Sklar	President	The Stella Group, Inc.	9 February 2006
Shalom Flank	Chief Technical Officer	Pareto Energy Limited	20 March 2006
Shawn Collins	Project Manager, Fuel Cells	United Technologies	3 March 2006
Shelly Strand	Marketing and Project Manager	Ecotope Consulting, Research, and Design	15 August 2005
Steffen Mueller	Senior Research Economist, Energy Resources Center	University of Illinois at Chicago	6 October 2005

(*Continued*)

APPENDIX 3.1 (Cont.)

Name	Title	Institution	Date
"Ted" (Edward C.) Fox	Interim Director, Energy and Engineering Sciences	Oak Ridge National Laboratory	1 August 2005
Thomas Grahame	Senior Research Analyst, Office of Fossil Fuels	U.S. Department of Energy	13 February 2006
Thomas Petersik	Former Renewable Energy Forecasting Expert	U.S. Energy Information Administration	1 February 2006
Tim Cross	Staff Writer	The Economist Magazine	30 September 2005
Toben Galvin	Senior Energy Analyst	Vermont Energy Investment Corporation	29 September 2005
Tom Casten	Chairman and Chief Executive Officer	Primary Energy	5 October 2005
Tommy Oliver	Assistant Director, Division of Economics and Finance	State Corporation Commission of Virginia	15 September 2005
Tommy Thompson	Energy Manager	Virginia Department of Mines, Minerals, and Energy	15 September 2005
Vijay	V. Vaitheeswaran	Energy Correspondent and Author	The Economist Magazine
6 March 2006			
Vikram Budhraja	Former President, Edison Technology Solutions	Edison International Company	27 September 2005
Wilson Prichett	Energy Consultant	ASEGI Incorporated	16 February 2006
Zia Haq	Operations Research Analyst	U.S. Energy Information Administration	9 February 2006

Notes

1 Solomon, B. D., and A. Banerjee. 2006. A Global Survey of Hydrogen Energy Research, Development, and Policy. Energy Policy 34: 781–792.
2 Shinnar, R., D. Shapira, and S. Zakai. 1981. Thermochemical and Hybrid Cycles for Hydrogen Production. A Differential Economic Comparison with Electrolysis. Industrial and Engineering Chemistry Process Design and Development 20 (4): 581–593.
3 Shinnar, R. 2003. The Hydrogen Economy, Fuel Cells, and Electric Cars. Technology in Society 25: 456–457.

4 International Energy Agency. 2004. Hydrogen and Fuel Cells: Review of National RandD Programs (Paris: Organization for Economic Cooperation and Development).

5 Moliner, R., M. J. Lázaro, and I. Suelves. 2016, 16 November. Analysis of the Strategies for Bridging the Gap Towards the Hydrogen Economy. International Journal of Hydrogen Energy 41 (43): 19500–19508.

6 Alanne, K., and S. Cao. 2017, May. Zero-Energy Hydrogen Economy (ZEH2E) for Buildings and Communities Including Personal Mobility. Renewable and Sustainable Energy Reviews 71: 697–711.

7 Moreno-Benito, M., P. Agnolucci, and L. G. Papageorgiou. 2017, 12 July. Towards a Sustainable Hydrogen Economy: Optimisation-Based Framework for Hydrogen Infrastructure Development. Computers & Chemical Engineering 102: 110–127.

8 Breeze, P. 2017. Chapter 7: The Hydrogen Economy. In Electricity Generation and the Environment, (Cambridge, MA: Academic Press), pp. 71–75.

9 Dagdougui, H. et al. 2018. Chapter 1: An Overview of Hydrogen Economy. In Hydrogen Infrastructure for Energy Applications, (Cambridge, MA: Academic Press), pp. 1–5.

10 Sovacool, B. K., and B. Brossmann. 2010, April. Symbolic Convergence and the Hydrogen Economy. Energy Policy 38 (4): 1999–2012.

11 Aronson, E. 2003. The Social Animal (New York: Worth Publishers, [Ninth Edition]).

12 Douglas, M. 2002. Purity and Danger: An Analysis of Concepts of Pollution and Taboo (New York: Routledge, 2nd Edition).

13 Berinstein, P. 2001. Alternative Energy: Facts, Statistics, and Issues (New York: Oryx Press).

14 Zubrin, R. 2007, Winter. The Hydrogen Hoax. The New Atlantis 15: 11.

15 Julio, F. S., and T. Homer-Dixon. 2004. Out of the Energy Box. Foreign Affairs 83 (6): 78.

16 European Commission Joint Research Center. 2004. Well-to-Wheels Analysis of Future Automotive Fuels and Powertrains in the European Context (Brussels: Joint Research Centre of the EU Commission).

17 American Physical Society. 2004. The Hydrogen Initiative (Washington, DC: American Physical Society).

18 Shinnar (2003: 456–457).

19 Romm, J. 2004, March 3. Testimony on Alternative Fuel Initiatives. Hearing before the House Science Committee: 27. p. 224.

20 Ogden, J. 2006, September. High Hopes for Hydrogen. Scientific American: 295: 94–101.

21 Ogden (2006: 97).

22 Hoffert, M. I. et al. 2002, November, 2002. Advanced Technology Paths to Global Climate Stability: Energy for a Greenhouse Planet. Science 298: 981–987.

23 National Commission on Energy Policy. 2004, June. The Car and Fuel of the Future: A Technology and Policy Overview (Washington, DC: NCEP).

24 National Research Council. 2004. The Hydrogen Economy (Washington, DC: National Academy Press).

25 Sperling, D., and J. Ogden. 2004, Spring. The Hope for Hydrogen. Issues in Science & Technology 20: 82–86.

26 Romm, J. 2007. Energy Myth Four—The Hydrogen Economy Is a Panacea to the Nation's Energy Problems. In Sovacool, B. K. and Brown, M. A. (Eds): Energy and American Society—Thirteen Myths (New York: Springer), p. 121.

27 Yacobucci, B. D., and A. E. Curtright. 2004, January 14. A Hydrogen Economy and Fuel Cells: An Overview. CRS Report For Congress.

28 Cohn, E. M. 1983. Energy in Wonderland—The Hydrogen Economy. Energy 8 (2): 30.

29 Geffen, C., J. Edmonds, and S. Kim. 2004. Transportation and Climate Change: The Potential for Hydrogen Systems. In Environmental Sustainability in the Mobility Industry (Detroit, MI: Society of Automotive Engineers).

30 McDowall, W., and M. Eames. 2006. Forecasts, Scenarios, Visions, Backcasts and Roadmaps to the Hydrogen Economy: A Review of the Hydrogen Futures Literature. Energy Policy 34: 1236–1250.

31 Wald, M. L., U.S. Drops Research into Fuel Cells for Cars, New York Times, May 7, 2009, available at www.nytimes.com/2009/05/08/science/earth/08energy.html

32 Merchant, B., Obama to Cut Funding for Hydrogen Fuel Cell Research, February 14, 2011, www.treehugger.com/corporate-responsibility/obama-to-cut-funding-for-hydrogen-fuel-cell-research.html

33 Schuelke-Leech, B.-A. 2014, April. Volatility in Federal Funding of Energy R&D. Energy Policy 67: 943–950.

34 American Institute of Physics, FY19 Budget Request: DOE Applied R&D Slashed Again, 2018, March 19, www.aip.org/fyi/2018/fy19-budget-request-doe-applied-rd-slashed-again

35 Sovacool and Brossmann (2010: 1999–2012).

36 Ale, B. B., and S. O. Bade Shrestha. 2008. Hydrogen Energy Potential of Nepal. International Journal of Hydrogen Energy 33: 4030–4039.

37 Árnason, B. A., and T. I. Sigfusson. 2000. Iceland: A Future Hydrogen Economy. International Journal of Hydrogen Energy 25: 389–394.

38 Balat, M. 2008. Potential Importance of Hydrogen as a Future Solution to Environmental and Transportation Problems. International Journal of Hydrogen Energy 33: 4013–4029.

39 Barreto, L., A. Makihara, and K. Riahi. 2003. The Hydrogen Economy in the 21st Century: A Sustainable Development Scenario. International Journal of Hydrogen Energy 28: 267–284.

40 Blanchette, S. 2008. A Hydrogen Economy and Its Impact on the World as We Know It. Energy Policy 36: 522–530.

41 Boudries, R., and R. Dizene. 2008. Potentialities of Hydrogen Production in Algeria. International Journal of Hydrogen Energy 33: 4476–4487.

42 Brey, J. et al. 2006. Designing a Gradual Transition to a Hydrogen Economy in Spain. Journal of Power Sources 159: 1231–1240.

43 Clark, W. W. 2008. The Green Hydrogen Paradigm Shift: Energy Generation for Stations to Vehicles. Utilities Policy 16: 117–129.

44 Clark, W. W., and J. Rifkin. 2006. A Green Hydrogen Economy. Energy Policy 34: 2630–2639.

45 Clark, W. W. et al. 2005. Hydrogen Energy Stations: Along the Roadside to the Hydrogen Economy. Utilities Policy 13: 41–50.

46 Crabtree, G. W., M. S. Dresselhaus, and M. V. Buchanan. 2004, December. The Hydrogen Economy. Physics Today 57: 39–44.

47 Doll, C., and M. Wietschel. 2008. Externalities of the Transport Sector and the Role of Hydrogen in a Sustainable Transport Vision. Energy Policy 36: 4069–4078.

48 Dunn, S. (2001). Hydrogen Futures: Toward a Sustainable Energy System (Washington, DC: Worldwatch), p. 157.

49 Dunn, S. 2002. Hydrogen Futures: Toward a Sustainable Energy System. International Journal of Hydrogen Energy 27: 235–264.

50 Dutton, A. G., and M. Page. 2007. The THESIS Model: An Assessment Tool for Transport and Energy Provision in the Hydrogen Economy. International Journal of Hydrogen Energy 32: 1638–1654.

51 Edwards, P. P., V. L. Kuznetsov, W. I. F. David, and N. P. Brandon. 2008. Hydrogen and Fuel Cells: Towards a Sustainable Energy Future. Energy Policy 36: 4356–4362.

52 Elam, C. C., C. E. Padro, G. Sandrock, A. Luzzi, P. Lindblad, and E. F. Hagen. 2003. Realizing the Hydrogen Future: The International Energy Agency's Efforts to Advance Hydrogen Energy Technologies. International Journal of Hydrogen Energy 23: 601–607.

53 Golstov, V. A., and T. N. Veziroglu. 2001. From Hydogen Economy to Hydrogen Civilization. International Journal of Hydrogen Economy 26: 909–915.

54 Golstov, V., T. Veziroglu, and L. Golstova. 2005. Hydrogen Civilization of the Future—A New Conception of the IAHE. International Journal of Hydrogen Energy 31: 153–159.

55 Greyvenstein, R., M. Correia, and W. Kriel. 2008. South Africa's Opportunity to Maximize the Role of Nuclear Power in a Global Hydrogen Economy. Nuclear Engineering and Design 238: 3031–3040.

56 Hodson, M., S. Marvin, and A. Hewitson. 2008. Constructing a Typology of H2 in Cities and Regions: An International Review. International Journal of Hydrogen Energy 33: 1619–1629.

57 Hotza, D., and J. C. Dinz Da Costa. 2008. Fuel Cells Development and Hydrogen Production from Renewable Resources in Brazil. International Journal of Hydrogen Energy 33: 4025–4035.

58 Lovins, A. B., and B. D. Williams (1999). A Strategy for the Hydrogen Transition. 10th Annual U.S. Hydrogen Meeting (Vienna, VA: National Hydrogen Association), pp. 1–17.

59 Maack, M. H., and J. B. Skulason. 2006. Implementing the Hydrogen Economy. Journal of Cleaner Production 14: 52–64.

60 Muradov, N. Z., and T. N. Veziroglu. 2005. From Hydrocarbon to Hydrogen-Carbon to Hydrogen Economy. International Journal of Hydrogen Energy 30: 225–237.

61 Muradov, N. Z., and T. N. Veziroglu. 2008. Green Path from Fossil-Based to Hydrogen Economy: An Overview of Carbon-Neutral Technologies. International Journal of Hydrogen Energy 33: 6804–6839.

62 Penner, S. 2006. Steps toward the Hydrogen Economy. Energy 31: 33–43.

63 Saxena, R. C. et al., 2008. Thermochemical Routes for Hydrogen Rich Gas from Biomass: A Review. Renewable and Sustainable Energy Reviews 12: 1909–1927.

64 Sherif, S. A., F. Barbir, and T. N. Veziroglu. 2005, July, 2005. Towards a Hydrogen Economy. Electricity Journal 18 (6): 62–74.

65 Taanman, M. et al. 2008. Fusion Paths for Micro Cogeneration Using Hydrogen in the Netherlands. Journal of Cleaner Production 16: 124–132.

66 Taljan, G. et al. 2008. Hydrogen Storage for Mixed Wind-Nuclear Power Plants in the Context of a Hydrogen Economy. International Journal of Hydrogen Energy 33: 4463–4475.

67 Tseng, P., J. Lee, and P. Friley. 2005. A Hydrogen Economy: Opportunities and Challenges. Energy 30: 2703–2720.

68 Winter, C.-J. 2004. The Hydrogen Energy Economy: An Address to the World Economic Forum. International Journal of Hydrogen Energy 29: 1095–1097.

69 Zidansek, A. et al. 2008, August 2009. Climate Changes, Biofuels, and the Sustainable Future. International Journal of Hydrogen Energy 34 (16): 6980–6983.

70 Aston, A. 2003, June 23. A Speed Bump on the Hydrogen Economy. Business Week 3838, p. 113.

71 Abraham, S. 2002. Remarks by Energy Secretary Spencer Abraham at the Global Forum on Personal Transportation Dearborn, Michigan November 12.

72 Berg, S. 2008, September 19. A Breakthrough in Hydrogen Research Could Help Curb U.S. Dependence on Imported Oil. Idaho Falls Post Register, p. A1.

73 Boehlert, S. 2003. Transition to a Hydrogen Economy: Opening Statement of Sherwood Boehlert, Congressman, House Science Committee (Washington, DC: Federal Document Clearing House Congressional Testimony).

74 Bush, G. W. (2003, February 6). Hydrogen Fuel Initiative Can Make a Fundamental Difference. Remarks on Energy Dependence at the National Building Museum.

75 Chandler, D. L., 2005, April 9. Fuel Cell Squeezes More from Petrol. New Scientist, p. 6

76 Clayton, M., 2004, February 19. One Step Closer to the Hydrogen Economy? Christian Science Monitor, p. 15

77 Saillant, R. B. (2002, November). Accelerating the Advent of the Hydrogen Economy. Electric Light and Power, p. 7.

78 Faulkner, D. L. 2005, July 27. Hydrogen Economy. Testimony before the Committee on House Government Reform Subcommittee on Energy and Resources.

79 Ledford, J., 2003, February 5. Hydrogen Economy Could Bring Radical Changes in What We Drive. Atlanta Journal-Constitution, p. 2B.

80 Lovatt-Smith, P. 2008, July 19. World with Less Oil. New Scientist, p. 20.

81 Menchaca, R. 2008, August 3. Hydrogen Research on Rise. The Post and Courier, p. A1.

82 Rifkin, J. 2003, February 23. The Forever Fuel. The Boston Globe, p. D12.

83 Rifkin, J. 2005, July 14. US Should Mount Large-Scale Effort to Establish a Hydrogen Economy. (R. Montagne, Interviewer) National Public Radio (NPR).

84 Roper, J. C. 2005. Heading to a Hydrogen Economy. The Houston Chronicle, June 24, p. B1

85 Editorial. 2005, April 24. Hydrogen Economy: Let's Put It on an Emergency Footing. The Santa Fe New Mexican (New Mexico), pp F-1.

86 Schwartz, P., and D. Randall (2003, April). How Hydrogen Can Save America. Retrieved August 15, 2007, from Wired.com:www.wired.com/wired/archive/11.04/hydrogen_pr.html

87 Spowers, R. 2003, January 13. The Forever Fuel. New Statesman.

88 Sovacool and Brossmann (2010: 1999–2012).

89 Freud, S. 1989. Introductory Lectures to Psychoanalysis (New York: Liveright Publishing, 2nd Edition).

90 Lacan, J. 1988. The Seminars of Jacques Lacan: Books I and II (Cambridge, MA: Cambridge University Press).

91 Evans, D. 1996. An Introductory Dictionary of Lacanian Psychoanalysis (New York: Routledge).

92 Nobus, D. 1998. Key Concepts of Lacanian Psychoanalysis (New York: Other Press).

93 Bales, R. F. 1970. Personality and Interpersonal Behavior (New York: Holt, Rinehart, and Winston).

94 Bormann, E. G. 1972. Fantasy and Rhetorical Vision: The Rhetorical Criticism of Social Reality. Quarterly Journal of Speech 58 (1972): 396–407.

95 Cragan, J. F., and D. C. Shields. 1981. Applied Communication Research: A Dramatistic Approach (Prospect Heights, IL: Waveland Press, Inc.).

96 Bormann, E. G. 1982a. Fantasy and Rhetorical Vision: Ten Years Later. Quarterly Journal of Speech 68: 288–305.

97 Bormann, E. G. 1985. Symbolic Convergence Theory: A Communication Formulation. Journal of Communication 35: 128–138.

98 Bormann, E. G., J. F. Cragan, and D. C. Shields. 1994. In Defense of Symbolic Convergence Theory: A Look at the Theory and Its Criticisms after Two Decades. Communication Theory 4 (4): 259–294.

99 Cragan, J. F., and D. C. Shields. 1995. Symbolic Theories in Applied Communication Research: Bormann. Burke. And Fisher (Creskill, NJ: Hampton Press and The Speech Communication Association).

100 Bormann, E. G., J. F. Cragan, and D. C. Shields. 2001. Three Decades of Developing, Grounding, and Using Symbolic Convergence Theory. Communication Yearbook 25: 271–313.

101 Shields, D. C. 2000, December. Symbolic Convergence Theory and Special Communication Theories: Sensing and Examining Dis/Enchantment with the Theoretical Robustness of Critical Autoethnography. Communication Monographs 67 (4): 392–421.

102 Csapo-Sweet, R. M., and D. C. Shields. 2000. Explicating the Saga Component of Symbolic Convergence Theory. Critical Studies in Media Communucation 17 (3): 318.

103 Bormann, E. G. 1982b. The Symbolic Convergence Theory of Communication: Applications and Implications for Teachers and Consultants. Journal of Applied Communication Research 10 (1): 50–61.

104 Shields (2000).

105 Stone, J. F. 2002. Using Symbolic Convergence Theory to Discern and Segment Motives for Enrolling in Professional Master's Degree Programs. Communication Quarterly 50 (2): 227–243.

106 Bormann (1982b).

107 Vasquez, G. 1994. A Homo Narrans Paradigm for Public Relations: Combining Bormann's Symbolic Convergence Theory and Grunig's Situational Theory of Publics. Journal of Public Relations 5: 203.

108 Foss, K. A., and S. W. Littlejohn. 1986. The Day After: Rhetorical Vision in an Ironic Frame. Critical Studies in Mass Communication 3 (3): 317–336.

109 Cragan, J. F., and D. C. Shields. 1992. The Use of Symbolic Convergence Theory in Corporate Strategic Planning: A Case Study. Journal of Applied Communication Research 20: 199–218.

110 Gunn, J. 2003. Refiguring Fantasy: Imagination and Its Decline in U.S. Rhetorical Studies. Quarterly Journal of Speech 89 (1): 41–59.

111 Bormann, E. G. 1977. Fetching Good Out of Evil: A Rhetorical Use of Calamity. Quarterly Journal of Speech 63: 130–139.

112 Scharf, B. F. 1986. Send in the Clowns: The Image of Psychiatry during the Hinckley Trial. Journal of Communication 37: 80–93.

113 Bormann (1982a).

114 Bormann, E. G., J. F. Cragan, and D. C. Shields. 1996. An Expansion of the Rhetorical Vision Component of the Symbolic Convergence Theory: The Cold War Paradigm Case. Communication Monographs 63 (1): 1–28.

115 Bormann (1972).

116 Bormann (1985).

117 Benoit, W. L. et al. 2001. A Fantasy Theme Analysis of Political Cartoons on the Clinton-Lewinsky-Starr Affair. Critical Studies in Media Communication 18 (4): 377–394.

118 Bormann et al. (2001).

119 Bishop, R. 2003. The World's Nicest Grown-Up: A Fantasy Theme Analysis of News Media Coverage of Fred Rogers. Journal of Communication 53: 16–31.

120 Chesebro, J. W. 1980. Paradoxical Views of Homosexuality in the Rhetoric of Social Scientists: A Fantasy Theme Analysis. Quarterly Journal of Speech 66: 127–139.

121 Cragan, J. F., and D. C. Shields. 1977. Foreign Policy Communication Dramas: How Mediated Rhetoric Played in Peoria in Campaign '76. Quarterly Journal of Speech 63: 274–289.

122 Drumheller, K. 2005. Millennial Dogma: A Fantasy Theme Analysis of the Millennial Generation's Uses and Gratifications of Religious Content Media. Journal of Communication & Religion 28: 47–70.

123 Duffy, M. E. 2003. Web of Hate: A Fantasy Theme Analysis of the Rhetorical Vision of Hate Groups Online. Journal of Communication Inquiry 27 (3): 291–312.

124 Barton, S. N. 1974. The Rhetoric of Rural Physicians Procurement Campaigns: An Application of Tavistock. Quarterly Journal of Speech 60: 144–154.

125 Shields, D. C. 1991. Drug Education and the Communication Curricula. In Decker, S. H., Rosenfeld, R. B., and Wright, R. (Eds): Drug and Alcohol Education across the Curriculum (Washington, DC: FIPSE, U.S. Department of Education), pp. 30–60.

126 Cragan and Shields (1992: 199–218).

127 Bush, R. R. 1981. Applied Q-Methodology: An Industry Perspective. In Cragan, J. F. and Shields, D. C. (Eds): Applied Communication Research: A Dramatistic Approach (Prospect Heights, IL: Waveland Press), pp. 367–371.

128 Vasquez (1994: 203).

129 Putnam, L. L., S. A. Van Hoeven, and C. A. Bullis. 1991. The Role of Rituals and Fantasy Themes in Teachers' Bargaining. Western Journal of Speech Communication 55: 85–103.

130 Dunn (2001).

131 Golstov and Veziroglu (2001).

132 Winter (2004: 1096).

133 Muradov and Veziroglu (2008: 6804).

134 Blanchette (2008: 522).

135 Clark et al. (2005: 44).

136 Clark and Rifkin (2006: 2634).

137 Golstov, Veziroglu, and Golstova (2005).

138 Berg (2008).

139 Blanchette (2008).

140 Greyvenstein et al. (2008).

141 Muradov and Veziroglu (2008).

142 Balat (2008).

143 Bush (2003).

144 Abraham (2002).

145 Sherif et al. (2005: 62).

146 Rifkin (2005).

147 Santé Fe New Mexican Editorial (2005).

148 Saillant (2002).

149 Schwartz and Randall (2003).

150 Schwartz and Randall (2003: 3).

151 Rifkin (2005: 1).

152 Dunn (2001: 12).

153 Schwartz and Randall (2003: 4).

154 Boehlert (2003: 2).

155 Hodson (2008: 1622).

156 Greyvenstein et al. (2008).

157 Hotza and Costa.

158 Edwards et al. (2008).

159 Rifkin (2003: D12).

160 Bockris (1999: 2).

161 Golstov and Veziroglu (2001: 911).

162 Dunn (2001: 27).

163 Elam et al. (2003: 602).

164 Editorial (2005: F1).

165 Menchaca (2008: A1).

166 Balat (2008: 142).

167 Edwards et al. (2008: 4361).

168 Muradov (2008: 6837).

169 Golstov, Veziroglu, and Golstova (2005: 155).

170 Balat (2008: 4014).

171 Crabtree, Dresselhaus, and Buchanan (2004: 39–44).

172 Eames, M. et al., 2006. Negotiating Contested Visions and Place-Specific Expectations of the Hydrogen Economy. Technology Analysis & Strategic Management 18: 361–374.
173 Barreto, Makihara, and Riahi (2003: 269-270).
174 Rifkin (2003: D12).
175 Spowers (2003: 34).
176 Eames et al. (2006).
177 Ale and Shresthab (2008).
178 Boudries and Dizene (2008).
179 Greyvenstein et al. (2008).
180 Hotza and Costa (2008).
181 Saxena et al. (2008).
182 Tannman et al. (2008).
183 Enevoldsen, P., B. K. Sovacool, and T. Tambo. 2014, December 12. Collaborate, Involve, or Defend? A Critical Stakeholder Assessment and Strategy for the Danish Hydrogen Electrolysis Industry. International Journal of Hydrogen Energy 39 (36): 20879–20887.
184 Andreasen, K. P., and B. K. Sovacool. 2014, November. Energy Sustainability, Stakeholder Conflicts, and the Future of Hydrogen in Denmark. Renewable & Sustainable Energy Reviews 39: 891–897.

4

TECHNOLOGICAL FRAMES

The interpretive flexibility of shale gas in Eastern Europe

For better or worse, shale gas has transformed the fossil fuel industry. What started off in a niche populated by "wildcatter" renegades of the energy sector has emerged into a multi–billion dollar global enterprise. Small companies have scaled up to industry size and so has the unconventional gas sector more generally. Shale gas— both the largest source and popular term for unconventional hydrocarbons— represents about 45 percent of total US gas output.[1] There, shale gas has become the new conventional source of energy. Soaring production has reversed a decade–long downward spiral and put the United States firmly on a path toward energy independence in natural gas. What is more, it has sent energy prices generally towards record low levels, which many observers regard as instrumental for boosting the reindustrialization of the American economy.[2,3]

One primary technical driver behind the "shale gas revolution" is a leap in technology innovation: hydraulic fracturing, or "fracking" for short. Combining horizontal drilling techniques with hydraulic fracturing, this advancement in technology allows exploiting reserves trapped in deep–rock formations. Fracking is an exemplary example of a contested novel technology, viewed by some as a cornucopia, by others as a curse.[4] On the one hand, it offers material benefits to the countries possessing and using the technology. These benefits come in the shape of economic welfare for local communities, tax dollar income for state coffers, or security gains in an energy world that turned more volatile. Potential risks stem from environmental harm caused by chemicals entailed in fracking fluids, methane migrating onto the surface, and the processing, storage or transport of contaminated flowback water.[5,6,7]

The contested nature of fracking on the one hand clearly relates to gains or losses that can empirically be measured, in terms of dollars on individual bank accounts, national import dependence ratios, or damaged acres of habitat. For societies facing the choice of embracing the novel technology, fracking in essence therefore amounts to a cost–benefit calculation. However, besides a purely empirical dimension, there are important contextual factors that matter. Technology innovation, the design of a technology, and its use in society are the result of a co–evolutionary process.[8,9,10] Complex dynamics between societal groups, the characteristics of the incumbent sociotechnical system and the societal interpretation of a technology all exert profound and prosaic influence over adoption preferences and pathways.[11,12,13,14,15,16] Yet so far most energy research centered on shale gas has focused only on the usual technical or economic concerns, ignoring more surreptitious yet salient social, political, and cultural elements. In short, it is the context and framing that gives empirics their meaning, but that process is rarely explored in the shale gas literature.

Based on earlier work[17] involving extensive field research and a set of 65 semi-structured research interviews, this chapter explores the "interpretive flexibility" or lack of "closure" of fracking in Bulgaria, Poland, and Romania. These countries, in line with Central Eastern Europe (CEE) more generally, face the choice of embracing shale gas as a potential revolutionizing domestic source of energy. This makes them an interesting case for studying how policy frames, narratives and discourses coalesce around a novel technology, and the inter-pretative frame countries and the communities that constitute them assign to it. This chapter explores the competing interpretive frames about shale gas within each country. Findings point to differing and indeed competing frames across Central Eastern Europe, and different sets of institutions sharing in those frames. These findings are not only interesting because CEE countries share strong similarities regarding external import dependence. They also provide for important insights on how the meaning of fracking is perpetually negotiated by those social groups connected to it.

Conceptualizing shale gas and hydraulic fracturing

Despite the claim that we are entering a new "shale gas revolution," its produc-tion does have a long history. In 1821, decades before the first oil well was drilled, commercial shale gas was extracted in Fredonia, New York. The United Kingdom's first well to encounter shale gas was drilled into the Upper Jurassic Kimmeridge Clay in 1875. In the Williston Basin in the United States, Bakken Shale has produced shale gas since 1953, and in 1969 the Atomic Energy Commission detonated an atomic bomb underground in Western Colorado to test "nuclear stimulation technology" that successfully "liberated natural gas that had been trapped in shale formations 7,000 feet deep."[18] In 1976, the U.S. Department of Energy initiated the Eastern Gas Shales Project at a cost of $200 million and the Mitchell Energy & Development Corporation (since merged with Devon Energy Corporation) started producing gas from the Barnett Shale of the Fort Worth Basin in 1981.[19,20] As the success of Mitchell Energy & Development became apparent, other companies aggressively began fracking so that by 2005, the Barnett Shale alone was producing almost half a trillion cubic feet per year of natural gas; the shale gas revolution had begun.

But for readers unfamiliar with shale gas production, this all begs the question: what exactly is it? Natural gas comes in a variety of hydrocarbon mixtures located in a variety of geological settings. What engineers call "wet gas" has a higher proportion of heavier molecules like ethane, propane, butane, and pentane, and comes in a liquid state; "dry gas" comes in gaseous state and (in the current market) fetches a lower price. Most "wet" and "dry" gas comes from "plays" (the industry term for "fields") composed of well-defined reservoirs with high rates of permeability. "Unconventional gas" refers to six types of gas plays where permeability is low: Coal-bed methane (contained in coal seams); tight gas (gas in low permeable formations); geo pressured gas (gas trapped in deep high-pressured reservoirs); gas hydrates (methane in the form of a crystalline solid that can be found in marine sediments); shale gas (gas trapped within shale formations of sedimentary rocks); and ultra-deep gas (often offshore reservoirs locked in high depths).[21,22,23]

The phrase "shale gas" therefore refers to natural gas extracted from gas shales, porous rocks that hold gas in pockets depicted in Figure 4.1. The key characteristic distinguishing shale gas from conventional gas is that it does not naturally flow into a well. It can be made to flow, however, by fracturing the formation containing the gas, artificially increasing its permeability. This is accomplished by "hydraulic fracturing," or "fracking" for short.[24]

Though the technology of shale gas production continues to improve, it involves at least seven elemental steps or phases:[25]

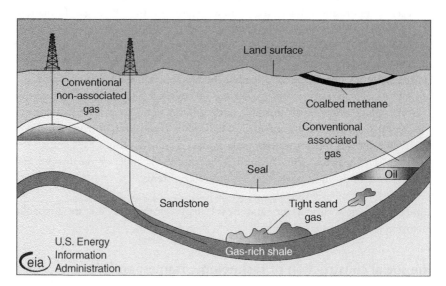

FIGURE 4.1 Major geologic sources of natural gas
Source: U.S. Energy Information Administration

- Seismic exploration refers to when underground rock formations are mapped using sound waves and three-dimensional reconstruction to identify the depth and thickness of appropriate shales. This may be done from the air, desktop (reanalysis of older data) or ground surveys;
- Pad construction refers to when a platform for a drilling rig is leveled and positioned over a discovered play, typically occupying about 5 acres (2 hectares);
- Vertical drilling refers to when a small drilling derrick drills as many as a dozen holes down to the shale rock, encasing a borehole in five concentric sleeves of steel and concrete near the surface, falling to one sleeve as the depth increases (suitable shales are typically 4,000–12,000 feet, or 1,220–3,660 meters, below the surface);
- Horizontal drilling refers to when a larger drilling derrick, 150 feet (46 meters) high, is assembled on site and slant-drills each well horizontally into the shale formation for thousands of feet in different directions, using gas sensors to ensure it stays within the seam;
- Fracking involves drilling, often at great depth, down into the shale layer, and then pumping water, sand, and chemicals into the shale at high pressure, releasing natural gas that flows back up with the drilling fluids. Hydraulic fracturing refers to when the concrete casing of the horizontal pipe is perforated with small explosive charges and water mixed with sand and other proppants is pumped through the holes at 5,000 psi (pounds per square inch) to fracture the rock with hairline cracks up to 1,000 feet (305 meters) from the pipe, taking 3 to 10 days for a single "frack" job;
- To access large shales or large amounts of gas, multi-stage fracking is typically utilized, where drillers not only use vertical wells but also horizontal ones, and they repeat the fracking process at each well as many as 20 times, with each pressurization fracturing a new region of the shale gas formation. Depending on the size of the well, the fracking process can involve injecting millions of gallons of water combined with thousands of gallons of proppants. Sustained production refers to when a "Christmas tree" valve assembly about the size of a garden shed, and a set of small tanks about the size of a small garage, remains on site to collect gas (and small quantities of oil), which then flows through underground pipes to a large compressor station serving a large number of wellheads and onwards to trunk pipelines;
- Waste disposal refers to when tanks collect water that flows back out of the well. The water is generally reused in future fracking, or desalinated and disposed of through sewage and wastewater systems.

Arguably the most important, and novel, parts of this process are hydraulic fracturing and horizontal drilling.

The social construction of technology, frames, and closure

To assess a technology as relatively new and complex as shale gas, I draw from literature arising from technological systems and the history of technology.[26,27,28,29]

This body of knowledge suggests that the evolutionary pathway of a novel technology (such as fracking) is not only a function of its technical qualities and characteristics but equally so of its perception within society. In this context, interpretative frames, and closure, are of great importance as they deal with how meaning is assigned to new technologies. One stream of thought known as the "social construction of technology," or SCOT, holds that technology emerges in society as a "seamless web"[30] or what Bruno Latour called a "sociotechnical imbroglio."[31] SCOT emphasizes the mutually constitutive nature of technology, usually described with the word technical or technological "artifact," and society.[32,33] Within this framework, four important elements have been developed: relevant social group, interpretive flexibility, technological frame, and closure and stabilization.

Relevant social group is used to denote the institutions and organizations that share the same set of meanings attached to a particular technology. Social groups that constitute parts of the "environment" for technology play a critical role in shaping and defining the problems that arise during the development of an artifact; social groups thus give meaning to technology, and define the problems facing that technology.[34] As Bijker notes, "an artifact does not suddenly leap into existence as the result of a momentous act by a heroic inventor; rather, it is gradually constructed or deconstructed in the social interactions of relevant social groups."[35]

Closely connected to the concept of a relevant social group is *interpretive flexibility*, which suggests that differing interpretations of technological artifacts are available. That is, different social groups see particular technologies in different ways. These technologies, then, become "heterogeneous" because their meaning, rather than being fixed, is interpreted and negotiated by those social groups connected to it.[36,37] Pinch and Bijker distinguish that technological artifacts possess interpretive flexibility at two levels: first, in how different social groups conceive of technology; second, that there is no one possible way that technologies are designed.[38] Hans Klein and Daniel Kleinman note that interpretive flexibility highlights that "technology design is an open process that can produce different outcomes depending on the social circumstances of development"[39] and artifacts are always the product of inter-group negotiation.

The idea of a *technological frame* or *interpretive frame* attempts to "frame" the interactions that occur between, rather than in or above, the actors. It comprises "all elements that influence the interactions within relevant social groups and lead to the attribution of meanings to technical artifacts—and thus to constituting technology."[40,41] Klein and Kleinman note that a frame can be defined as a:

> shared cognitive frame that defines a relevant social group and constitutes members common interpretation of an artifact. [...] A technological frame

can include goals, key problems, current theories, rules of thumb, testing procedures, and exemplary artifacts that, tacitly or explicitly, structure group members' thinking, problem solving, strategy formulation, and design activities.[42]

Making matters more complicated, interpretive frames have at least two additional features: heterogeneity and fluidity. Bijker comments that frames possess heterogeneity because technology does not belong exclusively to the cognitive or social domain. Components included are cultural values, goals, and theories, testing protocols and tacit knowledge but also hardware such as power plants, transmission lines, batteries, and cars. Second, technological flames are fluid because they are built up as part of the stabilization process of an artifact. Thus a technological frame does not reside in individuals or externally in the environment. Instead a "frame is largely external to any individual, yet wholly internal to the set of interacting individuals in a relevant social group."[43] Such frames can even become absorbed by national institutions or policies, with policymakers themselves often seeking to frame or reframe a technology or policy problem in ways more advantageous to their interests.[44,45,46] This can make frames both malleable and contested.

To illustrate both a diversity of potential frames and how they can become attached to fossil fuel infrastructure in particular, Table 4.1 emphasizes eight technological frames connected to two hydrocarbon pipelines in the Asia Pacific.[47] The Association of Southeast Asian Nations generally views the Trans-ASEAN Gas Pipeline Network as a mechanism to deliver natural gas to raise standards of living, Myanmar views it as a tool of political manipulation and control, Thailand as a way to gain strategic economic advantages over its neighbors, and Petronas as a component of its approach to economic development in Malaysia. Conversely, the World Bank Group generally views the Baku-Tbilisi-Ceyhan oil pipeline as a validation of the organization's commitment to sustainable development, British Petroleum and its affiliates as a way to transform the oil reserves of the Caspian Sea into profits, the European Union as a way to receive high quality oil without any of its negative downstream environmental impacts, and the government of Azerbaijan as a way to revitalize and remilitarize a country struggling since the collapse of the Soviet Union. There is no single frame of what either of these two pipelines is or ought to accomplish.

Closure and *stabilization* occur in technology when a consensus emerges that problems arising in the development of technology have been alleviated, and/or an agreement emerges concerning a dominant frame among relevant social groups. These problems need not to have been solved in the common sense of the word but only that the relevant social groups have perceived those problems as solved. Misa comments that closure has come to mean "the process by which facts or artifacts in a provisional state characterized by controversy are molded into a stable state characterized by

TABLE 4.1 Summary of the actors, interpretive frames, and interests associated with the Trans-ASEAN Gas Pipeline Network and the Baku-Tbilisi-Ceyhan Pipeline

Actor	Type	Pipeline	Interpretive frame	Underlying interest
Association of Southeast Asian Nations	International governmental organization	TAGP	Delivery system for natural gas	To use the natural gas flowing from the TAGP to equalize economic disparities between ASEAN members and other developed countries as well as fight poverty and grow economies within Southeast Asia
Myanmar	Government	TAGP	Tool of political control	To use natural gas revenues facilitated by the TAGP to purchase military weapons and maintain political authority
Thailand	Government	TAGP	Mechanism for competitive advantage	To transform Thailand into an "energy hub" and also accrue revenue for state-owned enterprises
Petronas	State-owned energy company	TAGP	Component of national identity	To promote a unique brand of Malaysian economic legitimacy
World Bank Group	Multilateral development bank	BTC	Validation of organizational strategy	To affirm the organization's style of development assistance and prove that extractive industrial projects can benefit communities and prevent environmental degradation
British Petroleum/ BTC Company	Multinational energy corporation	BTC	System of wealth transfer	To transform the "stranded" oil reserves of the Caspian Sea into billions of dollars of revenue and profits
European Union	Intergovernmental organization	BTC	Facilitator of environmental segregation	To "segregate" the environmental impacts associated with oil exploration, production, and distribution with use so that the European Union (and other importers) receive "clean" high quality crude oil while the environmental costs are distributed to Azerbaijan, Georgia, and Turkey

(Continued)

TABLE 4.1 (Cont.)

Actor	Type	Pipeline	Interpretive frame	Underlying interest
Azerbaijan	Government	BTC	Harbinger of national revitalization	To assert Azerbaijani sovereignty and independence from Russia and demonstrate the country's technical, economic, and political sophistication

Source: Sovacool 2011: 610–620.[48] Note: ASEAN=Association of Southeast Asian Nations. TAGP=Trans-ASEAN Gas Pipeline Network. BTC=Baku-Tbilisi-Ceyhan Pipeline

consensus."[49] In their examination of scientific controversies, Engelhardt and Caplan identify five different types of closure. Sound-argument refers to when closure occurs through correct and informed reasoning; consensus closure occurs when stakeholders agree it has been reached; procedural closure occurs when arbitration procedures declare it so; negotiation closure occurs when parties eventually reach a compromise; and natural-death closure occurs simply by boredom.[50]

Beder builds on this work and suggests that fives types of closure exist. Closure through:

- *Loss of interest* implies that a controversy ends because participants lose interest. Although no resolution or consensus has been reached, the issue has lost its importance or is no longer the focus of interest or controversy;
- *Force* means that a controversy is ended, although there is no rational basis for its resolution. It may occur when an external authority declares and imposes a decision, or by the use of state power, or even by loss of financial support;
- *Sound argument* is said to occur when participants freely concur that a particular solution is the most appropriate because they agree on the "facts" (based on an "epistemic" rather than social or political considerations);
- *Negotiation* often involves procedures agreed to by disputing parties for reaching a conclusion which may represent a compromise for them;
- *Consensus* implies that there is agreement because the force of one position has overwhelmed all others. It does not matter whether a certain point of view is well defended. It does not matter whether a correct or fair position has been reached or if a permanent or definitive solution has been found.[51]

Beder suggests that two forms of closure—negotiation and consensus—particularly depend on social dynamics and the interaction of participants. Also, while the term rhetorical closure implies a degree of stability, it is not always permanent: controversies and conflicts can re-emerge and new challenges can arise.

For any given technology or socio-technical system, closure or stabilization occurs when and if a social group and an artifact meld together; thus closure mechanisms can stabilize social groups as well as artifacts. This can include rhetorical closure, when a social group perceives a problem as solved, or closure through redefinition, when an artifact stabilized incompletely by one social group is stabilized more by association with a larger or more powerful social group. Bijker and Law emphasize that:

> Technology is stabilized if and only if the heterogeneous relations in which it is implicated, and of which it forms a part, are themselves stabilized. In general, then, if technologies are stabilized, this is because the network of relations in which they are involved—together with the various strategies that drive and give shape to the network—reach some kind of accommodation.[52]

Consequently, the concepts of closure and stabilization highlight that different interpretations of technology can lead to conflict and controversy, and the design process of technology will inevitably continue until such differences are resolved.

Moreover, exploring rhetorical closure can serve educational purposes at revealing different frames about technology and also exposing the broader cultural, economic, and political forces behind them.[53,54] Brown et al. add that:

> The orchestration of futures by powerful groups suggests that, perhaps, the future is not so uncertain and unpredictable as some of the other themes described above. It is as important then to recognise the need to strike a balance in my analysis between interrogating how futures are constructed and the openness and contingency this involves, and who constructs the future, and the relative closure that this involves.[55]

In this manner, examining closure can also reveal moments of leveraging the future, of "opening up" narratives so opportunities for the "the orchestration of opportunity" can be identified.[56]

Generally, the literature from SCOT suggests that frames and closure (or lack thereof) can exist in three configurations. In some instances, no frame may be present, as in the early history of the bicycle when no single dominant group had a set of vested interests in the success of the technology. Here, success depended on the formation of a constituency, and the technology is open-ended. In others, one frame may be present because a dominant group is able to insist upon its definition of the problems and solutions. This would be when closure is reached. A third configuration relates to when multiple frames are contested, and criteria external to such frames are needed to resolve differences. Shale gas in Eastern Europe, as we will see, exhibits the third type, as it is a relatively mature technology (with more than 20 years of operating experience) and yet faces contestation from (mostly environmental) opponents.

Research design and empirical strategy

Central Eastern Europe emerges as a crucial case for investigating technological frames in fracking. The region is highly energy import dependent, with some countries sourcing up to 100 percent of consumed natural gas from external suppliers. This supplier typically is Russia's Gazprom, a company that holds a dominant position in CEE markets. Among the countries studied in this chapter, Poland imports around 60 percent of consumption from Russia, Bulgaria 96 percent, and Romania 15 percent.[57] Romania's conventional production is set to decline significantly by 2020,[58] which makes the country's persistently low import ratio essentially a function of faltering domestic demand, a consequence of both a staggering economy and some (cautious) moves toward renewables and a deregulated domestic gas market. In all countries, coal plays an important role in domestic electricity production. Therefore, EU decarbonisation policies will force these countries to replace coal by less polluting sources, with gas resuming the role of a "bridge fuel."

Against this backdrop, the prospects of domestic natural gas production gain traction in policy debates.[59] It is particularly the region's shale gas reserves that have emerged as a focal point. The region as a whole is assumed to sit on vast unconventional hydrocarbon deposits. According to estimates by the US Energy Information Administration, Poland's 148 trillion cubic feet (Tcf) or 4.191 billion cubic meters (bcm) of technically recoverable shale gas reserves could roughly cover 250 years of current consumption. Bulgaria and Romania together hold 37 Tcf (1.047 bcm) or around 70 years of cumulative consumption.[60] The optionality of fracking would therefore benefit a set of countries that all come with similar external characteristics in the shape of a high import dependence on Russia and the decarbonization imperatives stemming from EU climate policies. Shale gas in CEE therefore provides for an ideal case to study the interpretative meaning assigned to the fracking technology by key stakeholders.

To identify specific frames connected to shale gas in Eastern Europe, the chapter relies primarily on original data drawn from 65 semi–structured interviews conducted by my colleague Andreas Goldthau in Bulgaria, Poland and Romania.[61] The bulk of interviews was conducted in Poland (27) and Bulgaria (27), whereas the remaining 9 interviews occurred in Romania. Interviews covered many relevant policy levels (national, regional, local) and actor groups (government, public and private companies, regulators, civil society) pertaining to shale gas.

Prior to conducting *sur place* interviews, a number of broader patterns characterizing each country's shale gas debate were identified through an assessment of media coverage and desktop research including peer-reviewed as well as more general sources of literature. Interview questions were then specifically designed to explore in more detail the dominant perceptions about shale gas and fracking among actors, to consolidate them in the shape of specific frames, and to unveil the importance assigned to each individual frame. Questions aimed at general perceptions, for instance, asked "What is the importance of shale gas to your country?" or "What is the public opinion on shale gas in your country/county/town, and has this changed?," whereas individual frames were assessed through questions such as "What are the financial considerations (e.g. tax, jobs, local benefits) for accepting or rejecting shale gas extraction in the country/county/town?" or "Could there be sufficient quantities extracted to impact the market price and ensure higher security of supply?" or "How can the environmental concerns, held by opponents of shale gas, be addressed?" Interview proceedings were kept open to ensure comments were adequately captured.

This dataset comes with three caveats. First, due to the sensitivity of the subject, a number of actor groups could not be included in the sample. This pertains primarily to state owned companies in Bulgaria. In addition, some interviews had to be held anonymously. Throughout the chapter, names are therefore omitted and interviewees are identified by their function and institutional background. Second, the sample is not fully representative regarding the entire spectrum of relevant societal groups and institutions. This caveat pertains primarily to civil society, which was covered but remains slightly under–represented. To the extent possible, this problem was addressed through triangulation. Third, as debates on fracking are still ongoing, interpretative frames remain in flux still. Data, therefore, capture the meaning given to fracking as a novel technology only between 2010 and 2014. The outlier here is Bulgaria, a country where "closure" could be observed in the discursive process, in the shape of a 2012 ban of shale gas, which was backed by a broad set of societal groups.

The chapter proceeds in two steps: first, it identifies dominant patterns of framing within each country; second, it compares variation in frames across countries, and assesses their relative dominance. The aim is to uncover whether dominant frames are similar or divergent across countries, and whether similar or divergent sets of actors and institutions endorse these frames.

Poland: "economic opportunity," "national security" and "environmental boon"

Poland has long been viewed as the front runner in shale gas exploration and possible exploitation throughout CEE. The country has been the target of major international energy companies, including ExxonMobil, Shell, ENI and Chevron, as well as companies specializing in shale gas, such as Talisman, Marathon, Cuadrilla and Lane Energy. The prospects of an unconventional gas sector scaling up in the country have become bleaker after most larger companies have turned their back on Poland after disappointing test drillings.[62,63] It is smaller companies that retain a presence in Poland.

The data on Poland reveal three specific and dominant interpretative frames for shale gas. The first, the "economic opportunity" frame, embraces shale gas as a source of economic welfare for the country. It is the positive effects on job creation, on revenues for state coffers at national and subnational levels, and on the competitiveness of the manufacturing industry that constitute the core of this frame. The "economic opportunity" narrative essentially mirrors statements made by Radoslav Sikorski, Poland's former foreign minister, to the effect that shale gas could make the country "a second Norway."[64] It is a frame that is shared by essentially the entire business community and state administration in Poland.

In the words of a director level representative of the Polish Confederation of Private Employers (PKPP Lewiatan), shale gas "can be important drivers for the Polish economy [because it] can produce cheaper gas for other economic sectors [such as] the chemical industry." In the same vein, a former advisor to the Polish Foreign Minister stressed lower energy costs and a "knock–on effect on energy consuming industry." Adding to this statement, a member of the Economic Policy Department of the Polish Ministry of Foreign Affairs pointed specifically to the economic benefits for the local communities in Poland, for whom "shale gas [is] a tremendous opportunity to speed up their development." Therefore, the Lewiatan representative also stressed, the Polish political elite is aware that "everything should be done to start a program quickly." An interviewee representing the Office of the Minister of the Treasury—the body overseeing state owned companies—in this context noted that "[our job] is to encourage investors to bring money here and drill and drill. [...] We will do our best to help investors, because our companies will benefit as well." This view was echoed by an Advisor to the Minister of the Environment, the ministry in charge of environmental oversight, who admitted that "[...] we know [about] the financial risks on the side of the companies, so we will try to regulate it in a way that will boost the level of investments and security of investments."

The "economic opportunity" frame was also fully embraced by the extractive industry. Although regulatory risk was mentioned as a significant impediment, "[s]hale gas can be an impulse and driver for the economy," according to a manager at United Oilfield Services, a private Polish energy service company, particularly against the backdrop of Polish gas prices generally perceived as being among the highest in Europe. Because of the potential of the Polish shale reserves, a former advisor to the Polish Foreign Minister alleged, "[s]ome of the [major international oil and gas companies] are viewing Poland as a base for their European oil and gas business." But also representatives of state owned companies point to the opportunities coming with shale gas—possibly a function of the Polish government insisting on bringing them into private industry–led gas exploration. As the chief economist of Polish national energy company Orlen insisted, "[i]t is not the case that the government pushes us into this. [...] We go in because we believe in it and put money on our bet that we will find gas." Benefits as perceived by state owned companies include foreign direct investment, gains in skills and expertise and technology transfer. Overall, the "economic opportunity" frame was most dominant among business, the oil and gas sector and the state administration, but it was also supported by a broader cross–section of societal actors.

The second dominant interpretative frame surrounding Polish shale centers on national security gains. This frame comes against the specter of Poland's historical experience of being geographically located between major powers, and particularly the trauma of 1939 when the country was partitioned by Soviet Russia and Nazi Germany. Reliable energy supply, notably with a view to Poland's strong reliance on Russian gas imports, forms an integral part of this "national security" frame. This frame is embraced by the bulk of actor sets represented in the interview sample.

As coined by the manager at United Oilfield Services, "energy is a foreign policy tool for Russia [and] shale gas opens up the possibility of being more secure from Russia's monopolistic position," which comes against memories of "a painful history for Poland under Russia." Other representatives of the Polish business community seconded this statement. Against this backdrop, "shale gas [is] part of [Poland's] diversification policy" (communication with representative of the Office of the Minister of the Treasury). Interviews with members of scientific community also revealed that this narrative resonates with academia to the effect that shale gas production "would be a milestone to be independent from Russia" (communication with professor of Polish Institute of Soil Science). Though not supporting shale gas as a key means to end Gazprom's dominance on the Polish market, the Chairman of Poland's Green Party admits that "[c]itizens see the opportunity to be independent […] from Russia." Reiterating this in a more nuanced way, it is the prospects of "stable supply of gas in Poland [without] any political strings attached" that motivate shale gas related policies on ministerial level, according to a member of Economic Policy Department of the Polish Ministry of Foreign Affairs. Even for Polish environmental NGOs, the aspect of "sovereignty is important," as alleged by a representative of Cleantech Poland, a consultancy firm active in the unconventional energy sector. That said, some environmental NGOs remained among the more cautious observers, hinting that the business community may only utilize the security narrative for fostering the shale gas cause. The reality, as a representative of Climate Coalition argued, is that "[w]e should not expect too much in the way of shale gas development in Poland. As a threat or a success story."

The third dominant frame pertaining to Polish shale gas is more contested, and it contrasts with the only negative frame to be found in the country. To some, fracking technology has been challenged as an "environmental bane" in Poland, as in other European countries and in some American states holding unconventional hydrocarbon reserves. The prospects of drilling using vertical hydraulic fracturing have indeed raised protests among local population. A prominent case in point is the village of Zurawlow, whose inhabitants have successfully mobilized against drilling activities planned by Chevron.[65] Skepticism has grown mainly over environmental concerns, notably groundwater safety issues and risks posed to local habitat. Environmental NGOs have vocally made the negative side effects of shale gas exploration and production a topic in public debates, which led to attempts of the Polish government to legally curb the ability of NGOs to get involved in shale gas matters.[66]

However, despite the presence of this negative frame, it is countered by another frame of "environmental boon." Contrary to a negative attitude toward fracking among some local communities and the environmental movement in the country, there is a consensus among most societal stakeholders that Poland urgently needs to decarbonize its energy system. As representatives of the employers' association, oil and gas companies and the business communities stressed during the interviews, Poland may face challenges related to the ETS, the European carbon trading scheme, and rising carbon prices—a function of the country's power supply relying primarily on coal. This may put economic pressure also on the economy. Shale was therefore generally supported among constituents in business and industry. More interestingly, however, it was also the environmental community that cautiously embraced unconventional gas as an opportunity to decarbonize the country in the long term. As a representative of Climate Coalition, a Polish environmental NGO, summarized:

> [s]hale gas is not a priority for our organization, [as] we look at the whole energy mix. [...] We believe [a] local use of shale gas [using] best available technologies could be a transition fuel that could complement the use of renewables [...].

Polish academics frequently also referred to gas and shale gas a "bridging fuel for renewables" (scientist at the Polish Institute for Sustainable Development and member of a county level climate project). Even the Green Party does not fundamentally go against shale gas. As its chairman indicated, his party was divided in two groups, with one opposing shale but the other trying to demonstrate that alternatives may neutralize the need to explore it at first place. Some environmental groups, such as the Environmental Protection League, would not even adopt a position on shale gas (communication with the President of the Board).

Even members of the energy sector confirmed during the interviews that Polish environmental NGOs are not principally against shale but stress that extraction should be carefully monitored (communication with corporate affairs manager, Talisman Energy). In sum, "[w]ithin environmental organizations, there is a pragmatic understanding" (communication with consultancy firm Cleantech Poland).

In all, natural gas, and also shale gas, is primarily viewed as a fuel that could lead the country into a low carbon future without putting the economy at risk. Put differently, whilst the "economic opportunity" frame would point to shale gas as a price hedge against EU decarbonisation targets, the lesser evil narrative stresses the element of a time hedge that's required to lower the country's carbon footprint. This frame is supported by members of the business community, parts of state administration and, surprisingly, even the environmental community.

Bulgaria: "economic sellout," "authoritarianism" and "environmental bane"

Bulgaria, by contrast to Poland, enacted a ban on fracking technology in 2012, which effectively stalled further exploration activities in unconventional gas in the country.[67] Chevron, the only foreign company active in shale gas in Bulgaria, has left the country.

Much as did Poland's political elite, Bulgaria's leadership sought to frame shale gas as a matter of national security and economic opportunity. The national security argument comes on the coattails of the January 2009 dispute between Ukraine and Russia, which led to a week–long gas cutoff and left the country out in the cold. It was the experience that one supply source poses a veritable security risk for the country which motivated statements by a former member of parliament and chairman of the Energy Independence Movement to the effect that "[s]hale gas is not only an industry, it's geopolitics." Moreover, rising energy prices have repeatedly led to protest in Bulgaria, the EU's poorest country, and even forced governments to leave office.[68] It has been alleged by Bulgarian authorities that domestic sources of gas would ameliorate the situation and reduce exposure against price hikes induced by oil indexation.[69] Moreover, as the former Deputy Minister of Economy, Energy and Tourism highlighted in an interview, the country's envisaged "reindustrialization process [requires] cheap gas," which would also "enter state budgets in the form of concession fees and royalties [and] fix our trade balance."

This national security and economic opportunity narrative resonated among other interviewees. For instance, as the chairman of the Bulgarian Federation of the Industrial Energy Consumers stressed, "there are two gas topics for Bulgaria. One is security of supply, the other is pricing." And yet, the security and economic benefit argument did not prove strong enough for a dominant interpretative frame to emerge and to be supported by key actors: Bulgaria banned shale gas exploration and the fracking technology in a parliamentary vote in January 2012.[70]

Among the various and competing negative interpretative frames in Bulgaria, three were identified as dominant. The first, the "economic sellout" frame, essentially labels shale gas as an attempt to exploit domestic resources for private gains. In this frame, foreign companies, rather than being perceived as engines of growth and sources of investment, are portrayed as the cause of unsustainable economic activity. As the co–chairman of the Bulgarian Green Party coined it, "foreign companies [...] do not create jobs for the local population [except for] guards, cleaning ladies and drivers," a statement that has been echoed by several local anti–shale activists. A leading activist of the Fracking Free Bulgaria Initiative further stressed that "investor is a dirty word in Bulgaria because for them it's easy to bribe officials and sign contracts with virtually no obligations." Public diplomatic attempts on part of the US administration and the American ambassador to convince the Bulgarian public and state officials of the positive effects of shale gas for the country were cited as (ironically) adding to resistance against the novel technology.[71]

The negative attitude toward foreign investors particularly in the extractive sector comes against the perception that "when things start to collapse, companies just give up but environmental problems remain" (communication with Member of "За Земята—Friends of the Earth Bulgaria"). As the Chairman of the Parliamentary Committee on Economic Policy, Energy and Tourism explained, "people [...] have seen in the past how [investors] caused damages and disappeared after that." What is more, municipalities do not perceive economic benefits arising from drilling activities. To the contrary, the mayor of the town of Toshevo stated, "[w]e don't have any gains for these 10 years [the lifetime of the lease], only a symbolic rent for the municipal lands." Concession fees for minerals are considered low by international standards, which has caused particular controversy in the case of the Bulgarian gold mining sector.[72] Whilst observers, notably from academia, stressed the opportunities that would be missed, the general perception that resonated among the scientific community was that "[t]here are no gains for the local population, only damages and problems" (communication with retired hydrology professor and anti–shale gas scientist).

Importantly, it is not only the perceived lack of economic opportunity and the exploitative element which emerged as part of this first dominant frame. The region of Dobrudzha, typically referred to as the "bread basket of Bulgaria," had been earmarked for exploratory drillings using the fracking technique. Potential pollution caused by shale gas extraction therefore entails the risk of depriving parts of the Bulgarian society of their economic base. Bulgaria's agricultural sector adds some 10 percent of GDP and accounts for 19 percent of total employment,[73] which compares to roughly 3 percent for the EU as a whole. Additionally, several interviewees stressed the importance of tourism as a key source of revenue for the region.

A final element of the "economic sellout" frame consists in the prevailing distrust pertaining to the motives of the government when pushing shale. The Chairman of the Parliamentary Committee on Economic Policy, Energy and Tourism hinted that "it looked like the government had an agreement with Chevron without [a proper] assessment of risks." This adds to a general perception among interviewees that the government adopted a deliberately soft touch approach toward shale gas investment in order to prioritize economic goals. According to the Chairman of the Bulgarian Federation of the Industrial Energy Consumers (BFIEC), the prevalent idea was that "the money will not go to people but to the state." This brings the shale gas industry close to allegations of state capture, which needs to be seen against the presence of high level corruption plaguing the country's energy sector.[74]

Overall, the "economic sellout" frame was most dominant among left and green parties and the local level public administration, but it was also supported by a broader cross–section of societal actors, ranging from business associations to academia.

The second frame that emerged made shale gas a case for promoting "energy authoritarianism" and lacking democratic and participatory governance in Bulgaria. This frame highlighted the problematic way the public interest was handled by the government and the energy industry—notably Chevron as the prime company active in unconventional gas. As numerous observers noted, information levels on shale gas among the public were generally low, and the former Deputy Minister of Economy, Energy and Tourism confirmed that "[t]he public lacked sufficient information and the government [...] and the business did not fill this gap on time. This is still the case—people still don't know what shale gas is and what the associated risks are."

Yet, interviews also revealed the perception that information on the fracking technology was strategically withheld by the government. Whilst an element of "secrecy" (member of Bulgarian Greens) characterized governmental activities relating to hydraulic fracturing and planned drilling activities, information campaigns were pursued "in a purely lobbyist fashion" (ibid.) and "[t]hings were presented as if everything is going to be alright" (communication with member of За Земята—Friends of the Earth Bulgaria). As a result, people started to source information from the internet, with *Gasland*, the American documentary on shale gas, being cited as a key source to gather detailed knowledge on fracking and its consequences. In this context, poor outreach toward key constituencies on behalf of the government emerged as a key element of the "democracy" frame. Dialogue between the government and citizens was described as poor or lacking. As observers noted, the communication between the private sector and civil society was channeled through the state administration, and even proponents of shale gas admitted that "Chevron [...] did not provide at that time any publicly available technological or other arguments regarding shale gas exploration" (communication with former Advisor to the Bulgarian Minister of Economy and Energy and former Bulgarian Ambassador at large for Energy & Climate Change).

In fact, municipalities and local populations typically learned about planned drilling activities only through official announcements. As one local mayor revealed, "[w]e don't have any communication with competent authorities, we only got a letter from the ministry when [Chevron] was granted a permit for exploration [...]." Protest leaders also indicated that local level referendums were blocked by GERB, the ruling party holding the majority in many municipalities (communication with business woman and protest leaders in Dobrich). This top–down and non–participatory approach on part of the government was perceived as "endangering democracy and the environment" (communication with leader of the anti–shale gas movement and co–chairman of Bulgarian Greens). Overall, this "energy authoritarianism" frame was shared by similar actors as the "economic sellout" frame but also found supporters among proponents of shale gas.

The third frame pertains, like in Poland, to shale gas being an "environmental bane." Actors embracing this frame on the one hand stress contingencies stemming from chemical substances used in fracking fluids. This line of argument pertains to a general fear that chemicals may harm natural habitats and cause environmental hazard. Arguably, this part of the environmental frame is partially a function of low levels of information among the population and stakeholders. As pointedly noted by two senior scientists of the Bulgarian Academy of Sciences' Geological Institute, "the population was frightened and waited for some kind of apocalypses provoked by shale gas and fracking," which was induced by "information that came from environmentalists' organizations—contamination of waters and soil with chemicals, radiation, earthquakes, and destructions." In fact, environmental groups emerged as important proponents of this frame. As the former Advisor of the Bulgarian Minister of Economy and Energy suggested, it was "predominantly young people who take Bulgarian environment and nature to heart," who drove the process, people "without famous faces [but] well educated" (co–chairman of Bulgarian Greens). Academics lamented during interviews that their scientific expertise was not heard. At the same time, however, there were clear rifts even within the scientific community, with some researchers siding with environmentalists and some not. Scientists critical toward fracking were alleged to form part of "epistemic communities [...] naturally leaning in favor of conventional technologies," rather than novel and unconventional ones (communication with members of the Board of Directors and head of the exploration unit of Oil and Gas Exploration and Production).

The "environmental bane" frame gained particular traction in the context of potential damage done to groundwater safety, and specifically an aquifer in the Dobrudzha region. While the risk of contamination was officially dismissed as "senseless theories" (former Deputy Minister of Economy, Energy and Tourism), protesters managed to establish a link to agriculture and food safety. This ensured anti-fracking support by the National Association of Grain Producers and among farmers (communication with Dobrich protest leader). Media, finally, tended to highlight environmental risks, too, which prompted comments from the energy business community to the effect that "[t]he media was not objective in covering the debate" (communication with member of the Board of Directors and head of the exploration unit of Oil and Gas Exploration and Production).

Overall, therefore, a broad coalition of actors supporting this third frame pertaining to the Bulgarian shale gas debate can be identified. It consists of left leaning political parties (notably the Greens and the Socialist Party), local municipalities and scientists but also industry associations and conservative parts of the society such as farmers. It is important to note in this context that the vast support of the Bulgarian anti–shale gas movement has triggered allegations of external influence in the shape of Russian money.[75,76] A comment of Bulgaria's former ambassador of Bulgaria in Russia, to the effect that "the environment [...] is also a very good 'business' for a lot of people" deliberately point to these allegations. Yet, whatever the strength of such potential external influence, my interviews primarily reveal broad support for the environmental risk frame among key groups within Bulgaria.

Romania: "economic opportunity," "environmental bane" and frame segregation

As the country covers most of its gas consumption through domestic production, shale gas does not play as vital a role in energy security. Romania thus offers an excellent opportunity to test perceptions and frames of shale gas in a country that is not yet adopting it.

In many respects, the Romanian shale gas tale features elements of both Poland and Bulgaria. On the one hand, the country's shale gas prospects have attracted the interest of international energy corporations such as Chevron (US), Sterling Resources (Canada), TransAtlantic Petroleum (Canada) or Hungary's MOL, which all acquired exploration and production licenses. Domestic state owned gas producer Romgaz has also expressed interest in joining shale gas exploration. Though the Social–Democrat Ponta government, an earlier critic of fracking, had put a temporary moratorium on shale gas exploration in 2012, the latter never went into effect. Overall, the country remains open to the novel technology and has not imposed a lasting ban on it as has neighboring Bulgaria.

On the other hand, fracking has become subject to fierce political debates between shale gas proponents and opponents. As observers have noted, growing social opposition against fracking may have been a decisive factor in making foreign companies let go of Romanian shale gas assets.[77] Officially a decision based on disappointing test drillings and a bearish international business environment, Chevron has announced it will leave Romania, whereas Sterling Resources sold their assets to Carlyle, the investor group.

As the interviews show, all frames identified for Poland and Bulgaria also exist in Romania, albeit with three important qualifiers: the "environmental boon" frame does not feature prominently; none of the frames is dominant among all actor groups; and frames seem much more attached to specific actor sets, with little overlap between them.

The "economic opportunity" frame, which was strongly pushed by the Ponta government despite rallying against shale gas whilst in opposition, mainly reso-nated among representatives of the state administration and independent observers. Cases in point are statements by the Romanian Ambassador–at–Large for Energy Security and Counselor to the Prime Minister to the effect that there is "a favorable case [to make] for shale gas exploration, due to the benefits it can bring to the national economy and energy security." Bringing on shale gas would also impact on the Romanian gas market structure, which by now is dominated by two major producers. "New competitors that would bring shale gas to the market, together with potential reserves in the Black Sea, would make competi-tion look totally different" (communication with energy analysts at Expert Forum, a Bucharest–based think–tank specializing in public policy and public governance reform). As stressed by various interviewees, there is particular potential for local development. Counties with potential for shale gas extraction typically are in the poorer parts of the country, and "badly [...] need better infrastructure—water, roads, scholarships for kids etc." (communication with the Director of the National Agency for Environmental Protection (ANPM), Ministry of Environment, Waters and Forests). Shale gas development therefore promises to enhance "the quality of life in the involved local communities, by creating jobs and raising local budget revenues" (Ambassador–at–Large for Energy Security).

Actors embracing the "economic opportunity" narrative partially overlap with those framing shale in terms of "national security." Particularly representatives of state institutions and the ruling party backed this frame. The general argument was that against the backdrop of falling domestic production, "shale gas [...] is a potential substitute and alternative to conventional hydrocarbons" (Member of Parliament, Industry and Services Commission). Shale gas "would be a contribution to our energy balance" and may "strengthen Romania's energy security [...], which is particularly important in the current regional geopolitical context" (Ambassador–at–Large for Energy Security). In short, given "Romania's energy security, we simply cannot ignore the potential contribution of natural gas from shale deposits" (President of the Romanian Agency for Mineral Resources (ANRM).

The two first frames stand in stark contrast to the narratives of "economic sellout" and "environmental bane," which are shared by similar actors sets—environmental organizations and think tanks. According to the President of Terra Mileniul III Foundation, an environmental NGO in Bucharest, little benefit is to be expected for the local population. Instead, shale gas exploration and production "will relocate oil workers for other parts of the country to the production site, while the more specialized employees will likely come from abroad [and] there are not many permanent positions in this industry." Seconding this, and echoing arguments made in the Bulgarian context, analysts at Expert Forum stress that "since the kind of job openings for the industry are mainly for the high skilled workers [there are no] opportunities for the locals—other than few low skills jobs." Furthermore, the royalty system was criticized as generating insignificant financial revenue. This is because royalties go to the central budget instead of local ones, and there is no mechanism to compensate municipalities for bearing the costs of shale operations. The perception of losing out economically coincides with fears among local communities pertaining to potential harm done to agricultural land and habitat. Some observers admitted that "[b]ringing more gas to the energy mix would replace coal [...]and bring down emissions in Romania" (analyst at Expert Forum). Others stressed Romania's longstanding history and sophisticated expertise in the oil and gas industry, and its ability to implement proper environmental oversight (communication with Members of Parliament, Industry and Services Commission). Still, it is environmental risks that dominate in that context. What is more, whilst the environmental risk narrative was most dominant at the national level of discourse, the economic sellout frame resonated strongest on the local level. As a consequence, "these two kinds of protest are augmenting each other" (analyst at Expert Forum).

The latter two frames extend into the "authoritarianism" narrative. It is mainly embraced by protest groups, NGOs and think tanks. This narrative on the one hand centers on institutional quality, notably the ability of state agencies to enforce laws pertaining to environmental oversight and corporate governance; their lack of coordination; and their ability to adequately deal with relevant policy aspects pertaining to the shale gas technology. In the words of an analyst at ExpertForum, "[...] the fickle institutions probably [are] one of the most important obstacles against public acceptance of fracking. The public does not trust institutions that today want to put a moratorium on fracking and tomorrow turn into enthusiastic supporters of it." Moreover, as various experts repeated, accessible scientific information in Romanian language is scarce, and the quality of information remained low. As a Counselor to the Minister for Energy, and Member of the Management Council of Transgaz mourned, "[d]iscussions in the media are mainly emotional, with little reference to science–based assessments of the costs and risks that shale gas operation can bring to Romania." *Gasland*, the American documentary was (again) frequently cited as a key source of information, which by some observers was deemed as "scientifically unfounded" (President of ANRM) or even "hostile [and] fueled by often radical and anarchistic movements" (Ambassador–at–Large for Energy Security).

This lack of institutional quality and transparency translates into distrust as a part of the "democratic" and participatory governance frame. This distrust comes in light of Romania's experience with an open–pit gold mining project at Roşia Montană in Alba county (Transylvania), a contested project for the use of the poisonous cyanidation mining techniques and allegations of corruption surrounding the permitting process.[78] It centers on the motivations of scientific inquiry and causes a deep divide between the "epistemic communities" (Radu Dudau, Director of Energy Policy Group, a Bucharest based think tank) of scientists and the business community on the one hand and the shale gas critics on the other. As the President of Terra Mileniul III alleged, "scientists and academics that often promote shale gas [...] are happy to get a small research contract financed by the oil companies." The respondent added that when coupled with "weak and indecisive [...] authorities [...] there is only the NGOs and the civil society left to counterbalance this."

In this respect, shale gas was also perceived as an issue of citizen participation. As the Director of ANPM admits, "[u]nfortunately, we started off on the wrong foot, due to inadequate communication." A Member of the Management Council of Transgaz, Romania's state-owned operator of the national gas transmission system went further by stressing that "[t]he fact that currently there is no participation of local communities to the prospective revenues makes them justifiably frustrated." Analysts at Expert Forum pointed to the heavy security apparatus deployed to protect shale gas exploration that raised "serious concerns about human rights infringement." In all, whilst the key actors embracing the "democratic and partici-patory governance" narrative are found among NGOs, civil society and think tanks, this frame also resonates among representatives of state institutions.

Conclusion

Traditional public policy experts and energy analysts often view energy systems such as shale gas or pipelines simply as energy delivery mechanisms that convert fuels into energy services, or transport fuels from one point to another. When more critical assessments occur, these are mostly about better managing the negative externalities from shale gas (e.g. by adopting the so called "Golden Rules of Gas" from the International Energy Agency[79]) or minimizing the risks of failure, accidents, and leakage.[80,81,82] Such assessments, while useful, rarely discuss issues relating to national development, equity, and visions of the future, nor do they show how the meaning for such systems can be contested.

Contrary to such analysis, the research conducted here of three Eastern European countries—Poland, Bulgaria, and Romania—suggests that shale gas can also be perceived, positively, as a way of accelerating economic development, serving national interests and increasing regional security, and encouraging less carbon intensive forms of energy supply. As Table 4.2 summarizes, however, other relevant social groups envisioned shale gas as an environmental bane due to its impacts on water and ecosystems, as a mechanism of transferring national wealth and assets out of the country, and as an energy system that only reinforces antidemocratic tendencies such as authoritative energy decision-making and the marginalization of the poor. Closure is elusive; contestation is significant. Assess-ments of shale gas that ignore these (sometimes hidden) social and political dimensions threaten to naturalize them as part of the normal environment and depoliticize them as acceptable risks.

Moreover, the fact that five of my frames—all but "environmental boon"—occur within relevant social groups across two countries or more implies a commonality of interpretation. That is, despite differing cultures, time periods, energy markets, modes of industrial cooperation, and national identity, there seem to be more universal frames that resonate with social groups at a deeper level. This may imply that the ability for an energy system such as shale gas to intersect positively with notions of economic growth and national security, or negatively with issues of equity and wealth or due process and anti-democratic processes.

In addition, certain sets of institutions tend to gather behind or share a certain frame. So it will, for instance, be civil society groups and municipalities that support an environmental frame in Bulgaria; in Poland, it is ministries, state energy companies and private businesses rallying behind a pro–jobs frame. So it is not necessarily one frame per institution but one frame that's shared by various sets of institutions. Sometimes one institution also buys into several frames at the same time (e.g. jobs and national security). The fact that many types of actors (such as business associations or public administrators) can populate multiple

TABLE 4.2 Summary of shale gas interpretive frames, locations, and relevant social groups in Eastern Europe

	Interpretive frame	Location	Relevant social group(s)	Description
Positive frames	Economic opportunity	Poland, Bulgaria, Romania	Government ministries, private energy companies	Shale gas will endow countries and communities with jobs, economic development, tax revenue, and in some situations a reindustrialization of the economy
	National security	Poland, Bulgaria, Romania	Energy consumers, some government ministries, private energy companies, members of civil society	Shale gas will enable countries to liberate themselves from dependence on energy imports (particularly from Russia) and enhance regional stability
	Environmental boon	Poland	Trade groups, unions, private companies, some environmentalists, some politicians	Shale gas will assist Poland in its process of national decarbonization
Negative frames	Environmental bane	Poland, Bulgaria	Local communities, environmental non-governmental organizations	Shale gas threatens water quality and availability, risks chemical pollution, and can accelerate species loss and the destruction of habitats
	Economic sellout	Bulgaria, Romania	Political parties, some civil society groups, some trade unions and business associations, local public administrators, some academics	Shale gas production merely transfers wealth and revenue out of domestic economies to foreign actors
	Energy authoritarianism	Bulgaria, Romania	Trade and business associations, local public administrators, some civil society groups	Shale gas decision-making is opaque, can concentrate political power, and marginalizes local communities

Source: Author, based on Goldthau, A., and B. K. Sovacool 2016, November[83]

frames at once may imply that they have learned about this universal appeal, and may switch between them depending on their audience.

That said, the fairly emergent nature of shale gas technology in Eastern Europe, and the admittedly contested nature of the visions, also implies that closure remains elusive. We do not see the loss of interest, appeal to sound argument, or consensus that Beder suggests would be needed for closure. Nor do we see successful negotiation (meanings attached with shale gas remain too divergent), or, in a positive light, force, for perhaps no actor has the necessary resources or authority yet to stabilize what shale gas means for all relevant social groups. Shale gas thus remains at a critical juncture where its meaning is co-evolving with the social groups and frames attached to it.

Finally, my study reveals that energy systems such as shale gas possess an interpretive flexibility and lack closure and stabilization, given that various social groups (stakeholders) continue to attach different, and at times conflicting, meanings to them. Shale gas, in other words, is poly-semiotic—it will provoke a heterogeneous mix of varying reactions based on both the type of relevant social group and the nature of the particular frame. In sum, the interpretive flexibility of shale gas reminds us that energy projects not only mark the physical landscape and contribute to the production and distribution of natural gas or electricity; they can also transfer what were once customary public resources into private hands, concentrate political power, become intertwined in national discourses of revitalization and strength, and validate distinct approaches to economic and social development.

These findings bear important implications for other fields of social inquiry, including security studies and political science—fields mostly concerned with the geopolitical or national and local political implications of shale gas. As this study reveals, new energy technologies such as shale gas interact in complicated ways with social, cultural, political, and economic forces. It is therefore the interpretative meaning attached to them that determines whether a novel technology will "succeed" or "fail." In other words, enhancing energy security by way of deploying novel energy technologies such as shale gas fracking is not simply a function of resource endowments and technological progress. Instead, it is the result of complex dynamics unfolding among social stakeholders and the related discursive processes that result. It is likely these dynamics, not hardnosed security imperatives by themselves, will eventually determine whether shale gas will go global—or not.

Notes

1 EIA. 2014. U.S. Crude Oil and Natural Gas Proved Reserves, 2013 (Washington, DC: US Department of Energy).
2 Reuters. 2013a, March 26. Shale Gas Lures Global Manufacturers to US Industrial Revival.
3 Washington Post. 2012, November 14. The New Boom: Shale Gas Fueling an American Industrial Revival.
4 Sovacool, B. K. 2014, September. Cornucopia or Curse? Reviewing the Costs and Benefits of Shale Gas Hydraulic Fracturing (Fracking). Renewable & Sustainable Energy Reviews 37: 249–264.

5 Gordalla, B. C., U. Ewers, and F. H. Frimmel. 2013. Hydraulic Fracturing: A Toxicological Threat for Groundwater and Drinking-Water? Environmental Earth Sciences 70 (8): 3875–3893.

6 Howarth, R. W., R. Santoro, and A. Ingraffea. 2011. Methane and the Greenhouse-Gas Footprint of Natural Gas from Shale Formations. Climatic Change 106 (4): 679–690.

7 Myers, T. 2012. Potential Contaminant Pathways from Hydraulically Fractured Shale to Aquifers. Groundwater 50 (6): 872–882.

8 Geels, F. W. 2002. Technological Transitions as Evolutionary Reconfiguration Processes: A Multi-Level Perspective and A Case-Study. Research Policy 31 (8–9): 1257–1274.

9 Geels, F. W., and J. W. Schot. 2007. Typology of Sociotechnical Transition Pathways. Research Policy 36: 399–417.

10 Schot, J. W., and F. W. Geels. 2008. Strategic Niche Management and Sustainable Innovation Journeys: Theory, Findings, Research Agenda, and Policy. Technology Analysis and Strategic Management 20 (5): 537–554.

11 Bijker, W., and J. Law. 1992. General Introduction. In Bijker, W. and Law, J. (Eds.): Shaping Technology/Building Society: Studies in Sociotechnical Change (Cambridge, MA: MIT Press), pp 1–29.

12 Coutard, O. 1999. Introduction: The Evolving Forms of Governance of Large Technical Systems. In Coutard, O. (Eds.): The Governance of Large Technical Systems (London: Routledge), pp. 1–16.

13 Hughes, T. P. 1987. The Evolution of Large Technological Systems. In Bijker, W. E., Hughes, T. P., and Pinch, T. J. (Eds.): The Social Construction of Technological Systems: New Directions in the Sociology and History of Technology (Cambridge, MA: MIT Press), pp. 51–82.

14 Misa, T. J. 1988. How Machines Make History, and How Historians (And Others) Help Them to Do So. Science, Technology, and Human Values 13 (3/4): 308–31.

15 Misa, T. J. 2003. The Compelling Tangle of Modernity and Technology. In Misa, T. J., Brey, P., and Feenberg, A. (Eds.): Modernity and Technology (Cambridge, MA: MIT Press), pp. 1–32.

16 Summerton, J. 1994. The Systems Approach to Technological Change. In Summerton, J. (Ed.): Changing Large Technical Systems (San Francisco, CA: Westview Press), pp. 1–21.

17 Goldthau, A., and B. K. Sovacool. 2016, November. Energy Technology, Politics, and Interpretive Frames: Shale Gas Fracking in Eastern Europe. Global Environmental Politics 16 (4): 50–69.

18 Jenner, S., and A. J. Lamadrid. 2013. Shale Gas vs. Coal: Policy Implications from Environmental Impact Comparisons of Shale Gas, Conventional Gas, and Coal on Air, Water, and Land in the United States. Energy Policy 53: 442–453.

19 Boyer, C. et al. 2011, Autumn. Shale Gas: A Global Resource. Oilfield Review 23 (3): 28–39.

20 Selley, R. C. 2012. UK Shale Gas: The Story so Far. Marine and Petroleum Geology 31: 100–109.

21 Wang, J., D. Ryan, and E. J. Anthony. 2011. Reducing the Greenhouse Gas Footprint of Shale Gas. Energy Policy 39: 8196–8199.

22 Jacoby, H. D., F. M. O'Sullivan, and S. Paltsev. 2012. The Influence of Shale Gas on U.S. Energy and Environmental Policy. Economic of Energy & Environmental Policy 1 (1): 37–52.

23 Bocora, J. 2012. Global Prospects for the Development of Unconventional Gas. Procedia—Social and Behavioral Sciences 65: 436–442.

24 Sovacool (2014: 249–264).

25 Ridley, M. 2011. The Shale Gas Shock (The Global Warming Policy Foundation GWPF Report).

26 Pinch, T. J., and W. Bijker. 1984. The Social Construction of Facts and Artifacts: Or How the Sociology of Science and the Sociology of Technology Might Benefit Each Other. Social Studies of Science 14 (3): 399–418.

27 Bijker, W. 1997. Of Bicycles, Bakelites, and Bulbs: Toward a Theory of Sociotechnical Change (Cambridge, MA: MIT Press).

28 Hughes, T. P. 1986. The Seamless Web: Technology, Science, Etcetera, Etcetera. Social Studies of Science 16 (2): 281–292.

29 Klein, H. E., and D. L. Kleinman. 2002. The Social Construction of Technology: Structural Considerations. Science, Technology, and Human Values 27 (1): 28–52.

30 Hughes (1986).

31 Latour, B. 1999. Pandora's Hope: Essays on the Reality of Science Studies (Cambridge, MA: Harvard University Press).

32 Bijker, W. 1993. Do Not Despair: There Is Life after Constructivism. Science, Technology, and Human Values 18 (1): 120–121.

33 Misa (1988).

34 Pinch, T. 1996. The Social Construction of Technology: A Review. In Fox, R. (Ed.): Technological Change: Methods and Themes in the History of Technology (New York: Harwood Academic Publishers), pp. 17–35.

35 Bijker (1993: 119).

36 Sovacool, B. K. 2011, May. The Interpretive Flexibility of Oil and Gas Pipelines: Case Studies from Southeast Asia and the Caspian Sea. Technological Forecasting & Social Change 78 (4): 610–620.

37 Noel, L., and B. K. Sovacool. 2016, July. Why Did Better Place Fail?: Range Anxiety, Interpretive Flexibility, and Electric Vehicle Promotion in Denmark and Israel. Energy Policy 94: 377–386.

38 Pinch and Bijker (1984).

39 Klein and Kleinman (2002: 3).

40 Bijker (1997: 123)

41 Law, J. 1991. Theory and Narrative in the History of Technology: Response. Technology and Culture 32 (2): 379–380.

42 Klein and Kleinman (2002: 31).

43 Bijker (1997: 123).

44 Kingdon, J. 1984. Agendas, Alternatives and Public (New York: Harper Collins).

45 Zahariadis, N. 2007. The Multiple Streams Framework. In Sabatier, P. (Ed.): Theories of the Policy Process (Cambridge, MA: Westview), 65–92.

46 Jones, M. D., and C. M. Radaelli. 2015. The Narrative Policy Framework: Child or Monster? Critical Policy Studies 9 (3): 339–355.

47 Sovacool (2011: 610–620).

48 Ibid.

49 Misa, T. J. 1992. Controversy and Closure in Technological Change: Constructing "Steel." In Bijker, W. and Law, J. (Eds.): Shaping Technology/Building Society: Studies in Sociotechnical Change (Cambridge, MA: MIT Press), p. 115–137.

50 Engelhardt, H. T. Jr., and A. L. Caplan (Eds.). 1987. Scientific Controversies: Case Studies in the Resolution and Closure of Disputes in Science and Technology (Cambridge, MA: Cambridge University Press).

51 Beder, S. 1991. Controversy and Closure: Sydney's Beaches in Crisis. Social Studies of Science 21: 223–56.

52 Bijker and Law (1992: 10).

53 Rood, C. 2017. Rhetorical Closure. Rhetoric Society Quarterly 47 (4): 313–334.

54 Pozzebon, M., R. Titah, and A. Pinsonneault. 2006. Combining Social Shaping of Technology and Communicative Action Theory for Understanding Rhetorical Closure in IT. Information Technology & People 19 (3): 244–271.

55 Brown, N., B. Rappert, and A. Webster. 2000. Introducing Contested Futures: From Looking into the Future to Looking at the Future. In Brown, N., Rappert, B., and

Webster, A. (Eds.): Contested Futures: A Sociology of Prospective Techno-Science (Aldershot; Burlington, VT: Ashgate), pp. 3–20.

56 Ibid., 13.

57 Eurogas. 2014. Statistical Report 2014 (Brussels).

58 KPMG. 2012. Central and Eastern European Shale Gas Outlook (London: KPMG Global Energy Institute).

59 Goldthau, A. 2012. Emerging Governance Challenges for Eurasian Gas Markets after the Shale Gas Revolution. In Kuzemko, C., Belyi, A., Goldthau, A., and Keating, M. (Eds.): Dynamics of Energy Governance in Europe and Russia (Basingstoke: Palgrave Macmillan), pp. 210–226.

60 EIA/ARI. 2013. EIA/ARI World Shale Gas and Shale Oil Resource Assessment. Technically Recoverable Shale Oil and Shale Gas Resources: An Assessment of 137 Shale Formations in 41 Countries outside the United States (Washington, DC: Department of Energy).

61 A huge thanks to Prof. Andreas Goldthau and his team for collecting this data and sharing it with me for this chapter.

62 FT. 2014a, January 15. Eni Joins Shale Gas Exodus from Poland.

63 Reuters. 2013b, May 8. Exit by Two Foreign Firms Leaves Polish Shale Gas under Cloud.

64 Kenarov, D. 2012, December 26. Poland's Shale Gas Dream. Foreign Policy.

65 The Guardian. 2015, January 12. Poland's Shale Gas Revolution Evaporates in Face of Environmental Protests.

66 Natural Gas Europe. 2013, April 1. Poland Proposes Restrictions to Shale Gas Opposition.

67 LaBelle, M., and A. Goldthau. 2014. The Governance of Shale Gas in Bulgaria. From Exploration to Bust. Oil, Gas and Energy Law Journal 12 (3): 1–38.

68 The Economist. 2013, February 15. Bulgaria's Electricity Prices. Protesting about Power Prices.

69 Reuters. 2010, July 15. Chevron Seeks to Explore for Shale Gas in Bulgaria.

70 The Guardian. 2012, February 14. Bulgaria Becomes Second State to Impose Ban on Shale-Gas Exploration.

71 Euractiv. 2012, February 7. US Tells Bulgaria Shale Gas Is Safe.

72 Kenarov, D. 2011, November 7. Where Your Gold Comes From: The Story of an Exploited Town in Bulgaria. The Atlantic.

73 European Commission. 2015. Agriculture and Rural Development (Bulgaria, Brussels: Member States Factsheets).

74 CSD. 2014. Energy Sector Governance and Energy (In)Se Curity in Bulgaria (Sofia: Center for the Study of Democracy).

75 FT. 2014b, November 30. Bulgarians See Russian Hand in Anti-Shale Protests.

76 FT. 2014c, June 19. Nato Claims Moscow Funding Anti-Fracking Groups.

77 Mihalache, A. E. 2015, March 30. No Shale Gas, after All—Implications of Chevron's Exit from Romania. Natural Gas Europe, pp. 1–2.

78 The Guardian. 2013, September 10. Romania Expected to Reject Gold Mine following Week of Protest.

79 IEA. 2012. Golden Rules for a Golden Age of Gas. World Energy Outlook Special Report on Unconventional Gas (Paris: OECD).

80 Alvarez, R. A. et al. 2012. Greater Focus Needed on Methane Leakage from Natural Gas Infrastructure. Proceedings of the National Academy of Sciences 109: 6435–6440.

81 Brandt, A. R. et al. 2014. Methane Leaks from North American Natural Gas Systems. Science 343 (6172): 733–735.

82 Sovacool, B. K., and V. Vivoda. 2014. Enhancing the Energy Security and Governance of Shale Gas. Oil Gas & Energy Law 12 (3): 1–35.

83 Goldthau and Sovacool (2016: 50–69).

5

DISCURSIVE COALITIONS

Contesting clean coal in South Africa

It was a strange request. Near the end of November 2006, South Africa's national state owned utility, Eskom, an organization not known for its appreciation of the arts and humanities, contacted a group of historians. Eskom and the Government of South Africa were struggling to come up with a better name to capture the essence of a bold new plan to build one of the largest coal-fired power stations in the world in Lephalale. Eskom's researchers had decided on "Project Alpha" for the $7.8 billion 4,800 MW coal-fired power plant, but everyone agreed this wasn't good enough. So, enter the historians, entrusted with finding a more suitable and culturally dynamic name. After discussing and deliberating came a better option: "Medupi." Medupi, an indigenous phrase, literally translates into something like "rain that soaks parched lands, giving prosperity."[1] As desired, the term linked the plans for the power plant with prosperity—in the form of energy supply—but unintentionally, to some of the pollution it could cause as well, given that acid rain, coal mining, and climate change could indeed "parch" the lands of Lephalale and the people living there.

This duality of interpretations in the naming of Medupi, most likely an accident, nonetheless embodies its wider, contested location in South African political language, and even in broader energy and climate discourses more generally. Plans for Medupi precipitated a fierce debate between government bureaucrats, industrial corporations and civil society. At the time of its conception, Medupi represented the state-owned utility Eskom's largest single investment in history,[2] and it is intended to generate 10 percent of South Africa's electricity. It would be (as of 2018) the world's ninth largest coal-fired power plant. It also reflects a state-of-the-art "clean coal" facility given its cooling systems, super-critical form of combustion, and pollution control.[3]

Yet even with its (admittedly contested) clean credentials, it would also add more emissions—30 million annual metric tons of CO_2-equivalent greenhouse gases[4,5]—than those of the 63 lowest-emitting countries combined. Moreover, Medupi's dependence on international financing—in particular, a substantial $3.75 billion loan from the World Bank, its first to South Africa since the end of apartheid roughly two decades ago—only heightened the controversy. Civil society organizations accused the government of "using blackmail and cheap tricks,"[6] while World Bank spokesman Sarwat Hussain retorted that "the campaign (by the NGOs) is based on half-truths."[7] This firestorm of rhetoric led the United States, Britain, the Netherlands, Italy and Norway to abstain from the World Bank vote. The controversy intensified and spilled over into widespread protests involving 65 South African civil society organizations and perhaps an even greater number of Western environmental groups.

How are we to better understand the discursive dynamics behind the Medupi controversy? Based on earlier work,[8] this chapter does so by viewing climate change and energy supply not as natural or technical phenomena but as constantly negotiated discourses. Based on a targeted sampling of project documents, reports, testimony, and popular articles, the study then maps the discursive justifications behind the project (economic development, environmental sustainability and energy security) as well as those against it (corruption or maldevelopment, environmental degradation, and energy poverty and racism).

Tracing the intricacies of the Medupi controversy does more than offer insight into the dynamics of energy policy and planning in South Africa. To be sure, it informs both the debate emerging over Kusile, the next coal-fired baseload plant in South Africa's approved build-program, as well as the country's broader, long-term Integrated Resource Plan, which at present appears unlikely to substantially reduce emissions,[9] improve energy security,[10] accelerate the uptake of large amounts of renewable energy,[11] or provide universal electricity access.[12] However, more broadly the polarizing dispute over Medupi is emblematic of future global struggles as to whether we ought to rely on cheap, abundant fossil fuels, or pivot globally to more renewable, less environmentally deleterious sources of energy supply. As global energy consumption doubles between 2015 and 2040, and triples by 2060, investment decisions in developing nations like South Africa will either lock in future emissions or reorient growth toward sustainability. It is "at this international carbon margin where the battle on global warming, and the planetary future, ultimately will be won or lost."[13] Closely looking at Medupi can shed light on the dramatic difficulty of this task, and inform future strategies for sustainable development.

Conceptualizing "clean coal"

The case for clean coal—by no means a given[14,15]—begins with the assumption that coal is and will remain a firmly entrenched part of the existing, and future, energy system. The world is already home to about 24,000 coal mines and the coal sector provides approximately seven million fulltime jobs.[16] Moreover, the coal sector is expected by some to grow faster than other fossil fuels in the near term. For instance, some major economies such as South Africa, the United Kingdom, Poland, and Australia rely on coal to generate 90 percent or more of their electricity, and the rapidly developing economies of China and India depend on coal for approximately 50 percent or more of electricity.[17]

Despite global alarm over greenhouse gas emissions and climate change, the International Energy Agency forecasts that coal could surpass oil as the world's primary source of energy in only a few years.[18] As Maria van der Hoeven, Executive Director of the IEA, put it when interpreting this data:

> Thanks to abundant supplies and insatiable demand for power from emerging markets, coal met nearly half of the rise in global energy demand during the first decade of the twenty-first century … In fact, the world will burn around 1.2 billion more tonnes of coal per year by 2017 compared to today—equivalent to the current coal consumption of Russia and the United States combined. Coal's share of the global energy mix continues to grow each year, and if no changes are made to current policies, coal will catch oil within a decade.[19]

For instance, there are approximately 1,200 coal-fired power plants constituting 1,400 GW of capacity in differing stages of planning and construction around the world, with significant growth expected in China, India, Russia, and the United States.[20] The International Energy Agency also predicted in 2008, when planning discussions for Medupi were underway, that 87 percent of new energy demand from 2006 to 2030 would be from non-OECD countries.[21] The implication at that time was not lost on South African planners: countries must use cheaper, dirtier fuels like oil and coal if necessary to supply energy and keep electricity and petrol costs low. Coal, in other words, was believed to still have an important role to play in the global (and South African) energy system.

"Clean coal" thus focuses on making this conventional and incumbent energy source as environmentally sustainable as possible. It is an umbrella term for processes and approaches that seek to minimize emissions of greenhouse gases and other pollutants from the use of coal for electricity and industry.[22] As defined by the U.S. Environmental Protection Agency, clean coal:

> means any technology, including technologies applied at the pre-combustion, combustion or post-combustion stage, at a new or existing facility which will achieve significant reductions in air emissions … associated with the utilization of coal in the generation of electricity, process steam, or industrial products.[23]

Advocates of this strategy point out that it is relatively easy for CO_2 to be sequestered from power plants and other emitting industries. Moreover, they highlight that clean coal offers a major technological "fix" to the problem of climate change, since they have the potential to stop atmospheric GHG concentrations from increasing.

For example, clean coal sponsors mention that best practices can greatly minimize any damage that might occur during the mining, transport, and combustion of coal. Already established control methods exist to minimize miner exposure to coal mine dust, the primary contributor to coal workers' pneumoconiosis, commonly known as black lung disease. These include applying water to lower dust from roads, using brushes to clean conveyor belts, and improving air ventilation and circulation.[24] Furthermore, Methane Control and Prediction (MCP) systems can be implemented to monitor fugitive emissions, and enable the recovery or capture of coalbed methane—which simultaneously enhances safety, reduces leakage, and provides a revenue source from the relatively clean natural gas byproduct.[25]

Downstream at power plants, the installation of flue gas scrubbing systems, baghouse filters, electrostatic precipitators, flue gas desulfurization systems, carbon injection systems to absorb mercury, and selective catalytic reduction equipment can mitigate as much as 90 percent of the emission of harmful air pollutants. Additionally, potentially hazardous releases of chemical pollutants to the air and water can be tracked, with adverse information given to local communities to minimize their exposure.[26]

Lastly, at the backend in terms of waste storage, CO_2 is already routinely captured as a by-product of ammonia and hydrogen production and from limestone calcinations. Moreover, engineers have been experimenting with pre-combustion capture, post-combustion capture, and oxyfuel combustion for decades.[27] The number of large-scale industrial facilities with carbon storage systems operational or under construction exceeded 75 in disparate locations such as Australia, China, Canada, Germany, Netherlands, Poland, and the United Kingdom. Sixteen of these projects are big enough to capture 36 million tons of carbon dioxide per year.[28]

Medupi—a massive endeavor shown in Figure 5.1 still being constructed as of 2018, expected for completion in 2020—was at one point the largest infrastructure project in the world when it had a peak of 18,000 construction workers and absorbed a significant share of the world's steel and concrete. Notwithstanding its mammoth size, Medupi can be legitimately classified as clean coal for three reasons. It would rely on "dry cooling" technology to minimize water use. It would burn coal at supercritical temperatures (improving thermal efficiency). And it would feature advanced fabric filters to reduce ash as well as flue gas desulphurization and low nitrogen oxide burners to control temperature and mitigate acid rain emissions.[29]

Climate change and energy security as "discursive coalitions"

This chapter examines clean coal through the lens of "discourse" and "discursive coalitions." A small albeit growing number of studies have recently begun to

FIGURE 5.1 The Medupi coal fired power station under construction in 2014
Source: Author

investigate the discourse surrounding climate change, energy security and sustainable development. However, these works frequently focus on the rhetoric of climate science within scientific communities, common lines of argument within alternative energy policy discussions, political commentary on climate change, and/or media representations of climate change. They rarely look at how discourses become institutionalized within particular stakeholders or groups of actors, usually study newspaper or television coverage but not deeper policy briefs and reports, tend to restrict their focus to "the public" or "scientists," and predominately look at case studies in Europe or North America.

By contrast, this chapter looks at the direct actors involved in emitting greenhouse gases (such as energy companies), designing policy (government administrators), providing loans (multilateral development banks), and lobbying for policy change (non-governmental advocacy groups) in the developing economy of South Africa. To say that it investigates climate change and energy security "discourses" does not imply that it assesses language or rhetoric independent of context. Instead, the chapter draws on the term as used by various critical theorists, especially Michel Foucault,[30] Mikhail M. Bakhtin,[31] Maarten Hajer,[32] and Arturo Escobar.[33]

Discourse, for these thinkers, refers to a "historically emergent collection of objects, concepts, and practices" that "mutually constitute" each other to cohere into stable meaning-systems. In this manner a discourse refers to a shared meaning of a phenomenon, a collective way of apprehending and comprehending the world.[34] Situated at an intersection of institutions, language, and power, discourses extend beyond the intentionality of mere individuals.[35] For Bakhtin, discourse is a dialogic stream of speech that merges with ideology and history to reflect a constant struggle between actors over who can impinge language with their own preferred meanings. As Bakhtin wrote:

> The word is not a material thing but rather the eternally mobile, fickle medium of dialogic interaction. It never gravitates toward a single consciousness or voice. The life of the word is contained in its transfer from one context to another, from one social collective to another, from one generation to another. In this process the word does not forget its own path and cannot completely free itself from the power of these concrete contexts into which it has entered.[36]

Thus, for Bakhtin (and others sharing his views), discourse is neither fixed nor formed of free floating signifiers, it instead is dynamically mediated through social processes and power structures.[37] Foucault similarly described this intersection as the ritualization of power into a truth regime, and contended that "each society has its regime of truth, its general politics of truth: that is, the types of discourse which it accepts and makes function as true; the mechanisms and instances which enable one to distinguish true and false statements."[38]

When applied to energy and climate topics, these truth regimes demarcate what is considered possible and represent a particular way of viewing reality. Discourse thus frames problems and their necessary solutions. As Karen T. Litfin put it:

> As determinants of what can and cannot be thought, discourses delimit the range of policy options, thereby functioning as precursors to policy outcomes ... The supreme power is the power to delineate the boundaries of thought—an attribute not so much of specific agents as it is of discursive practices.[39]

Discourses therefore serve to explain how dominant or even hegemonic practices and ideas come to be stabilized.

Viewing discourse in this way has significant implications for climate and energy practice and scholarship (and even consumption). Recognizing climate change and energy policy as discursive formations is to understand them as more than processes that occur to people—more than the mere emission of greenhouse gases, consumption of energy fuels and services, or construction of energy infrastructure. Climate and energy policy discourses do not simply describe human relationships with material artifacts like coal-fired power plants but actively structure these relationships by way of narratives and ideologies. Thus discourses do not merely float ephemerally like ghosts around various actors but instead are situated within specific organizations, crystallizing into practices and ways of reasoning.

Discourses in this sense often emerge as particular narratives with common themes. Escobar, for example, contends that institutionalized discourse in development projects draw various narratives together into singular "norms about development,"[40] a process that he terms "discursive homogenization."[41] Analyzing the discourse of a particular institution, then, requires examining the "argumentative structure in documents and other written or spoken statements" which provides insight into the interplay of language, identity, and policy.[42] Deconstructing such discourses can also highlight where reforms and improvements can be made. Or, as Escobar put it, "detailed studies of specific development practices (e.g. programs or institutions) are likely to contribute to this dismantling process by providing the context for modifying current practices."[43] In addition, examining the way conservationists situate their environmental advocacy in the context of competing, local economic claims—and moving beyond the ideological purity of conservationism, which Adams and McShane characterize as the "myth of wild Africa"[44]—is necessary to refine our understanding of the discourses of sustainable development. In essence, the chapter helps reveal the multidimensional nature of energy security and climate change, showing how various aspects of each conflict and complement each other.

Viewing discourse in this way implies that representations of different energy and climate problems can be contested, that their meaning is provisional rather than permanent. It also suggests that discourses will emerge as particular narratives with common themes, something Hajer called a "discursive coalition" which ties institutions and ideas together around particular shared conceptions of reality.[45] Escobar goes even further to argue that discourses can become "hegemonic" when they presume superiority and exclusivity in knowledge.[46] Institutionalized discourses frequently function as "knowledge industries," where "experts" generate coherent policy narratives and conceptualize, plan, and implement projects.[47] Their repeated performance both structures and constrains conditions for agency.[48] I embark on just such a mapping of argumentative structures involving the Medupi controversy.

Research design and empirical strategy

To identify the competing discourses behind the Medupi project, with my colleague William Rafey I isolated the major institutions and organizations involved in its financing, construction, operation, and ownership over the most significant period of its planning and development (2006 to 2010). I identified the most vocal supporters of Medupi within the Government of South Africa (GoSA) as the South African Department of Public Enterprises (DPE), the South African Department of Energy (DoE), the South African Department of Environmental Affairs (DEA), and Eskom. I also included among the project's advocates its two largest financiers, the World Bank and the African Development Bank (AfDB). Opposition to Medupi, in contrast, centered on three non-governmental organizations: the South African environmental justice advocacy group groundWork (also known as Friends of the

Earth South Africa), the Johannesburg-based Earthlife Africa Jhb (hereafter Earthlife), and the Washington, D.C.-based human rights organization Africa Action. I also included—as less vocal contributors to the anti-Medupi discourse, in terms of sheer number of documents—Greenpeace and the World Resources Institute (WRI), the Wildlife and Environment Society of South Africa (WESSA), and the South Durban Community Environmental Alliance.

The research process involved two central components: (1) the compilation and review of every available public document I could collect—press releases, speeches, publications, op-eds, and commissioned reports—concerning the Medupi project from its conception in 2006 to June 2010 written by the institutions and organizations above, totaling 88 primary documents identified in Table 5.1, and (2) a thorough search of relevant secondary domestic South African and international media literature over the period June 2009—June 2010 for public statements made by the selected actors.

I chose to include this preliminary survey of the media out of the recognition that the debate over Medupi is explicitly public and a number of perspectives were articulated and re-articulated in remarks made to the press. The period chosen for the review of secondary sources—the 10 months leading up to the World Bank decision and the two months after—is where the bulk of the discourse was concentrated. In the review, I examined the first 1,000 most relevant documents of 2,218 news media results via Eskom's news search engine for all documents containing "Medupi" over my selected time period (July 2009 to July 2010). I also searched Legalbrief Environmental, the South African news aggregator related to environmental and development law, for all documents containing Medupi, finding 59 relevant linked articles (July 2007 to July 2010). Lastly, I looked through all of the actors' self-reported media hits, recognizing the potential for each organization to have an interest in closely monitoring their own media coverage.

While South African media sources accounted for most of the literature I analyzed, all three of my search methods retrieved sufficient amounts of global news coverage such that my study included both national and international sources. Research was conducted exclusively in English, which did not appear to exclude any unique perspectives, even those from Afrikaans publications like *Beeld* or *Die Burger*.

Discursive coalitions supporting the Medupi clean coal facility

In particular, I identify three distinct discursive justifications supporting the Medupi coal plant: (1) a theme of economic development to assist South Africa's recovery from a global recession and revitalize its industrial competitiveness; (2) a theme of environmental sustainability emphasizing the project's place in a low-carbon future; and (3) a theme of energy security that supports access to electricity for all South Africans.

TABLE 5.1 Primary sources of data for "pro" and "anti" Medupi discourses

a. Top panel: Pro-Medupi discourses

Pro-Medupi institutions	World Bank	African Development Bank (AfDB)	Department of Public Enterprises (DPE)	Department of Energy (DoE)	Department of Environmental Affairs and Tourism (DEAT)	Eskom	Government of South Africa (GoSA)	Total unique documents
Official, primary documents	12	6	8	4	3	24	3	60
Secondary media hits	7		2	2	1	10		10

b. Bottom panel: Anti-Medupi discourses

Anti-Medupi groups	Africa Action	Earthlife Africa Jhb	Greenpeace	groundWork (Friends of the Earth South Africa)	Friends of the Earth International	South Durban Community Environmental Alliance	World Resources Institute (WRI)	World Wildlife Fund (WWF)	Wildlife and Environment Society of South Africa (WESSA)	Total unique documents
Official, primary documents	2	7	3	11	1	1	1	3	0	28
Secondary media hits	5	4	2	3	2	1	0	3	1	16

Source: Author, based on Rafey and Sovacool 2011: 1141–1151[49]

Economic development

The first theme championed by supporters of the Medupi power plant is economic development. Cheap, plentiful energy is framed as crucial to fuel continued economic growth, help South Africa recover from a global financial crisis, and bolster industrial competitiveness. As former Minister of Public Enterprises Barbara Hogan (May 2009 to October 2010) commented in her public defense of Medupi, "If they [the people of South Africa] did not have that power in their system, they could say goodbye to their economy and to their country."[50] In the eyes of its supporters, Medupi will have "a substantial macroeconomic footprint,"[51] "a major impact on the economy of South Africa,"[52] and "a large positive impact in the sub-region … creating the conditions for secure and predictable industrial and commercial growth."[53] The loan is "required for economic growth," deemed "necessary so that we do not derail the country's economic growth and development objectives" and a "statement of confidence in the economic outlook of the wider Southern African region, its people, and its prospects."[54]

Thus, the global economic recession that began in early 2008—and South Africa's coinciding rolling blackouts—vastly strengthened the hand of GoSA to justify its loan application. World Bank reports conclude that South Africa's "recovery will be thwarted by energy shortfalls unless these are addressed in a timely fashion" and that "[t]he proposed Project supports South Africa's strategic response to the impact of the global economic crisis."[55] Medupi's relatively fast scale-up of baseload power is necessary to "avoid black-outs and its consequent negative economic impact."[56] Blackouts are understood to implicate the wider sub-Saharan region, which depends on Eskom for more than 60 percent of its electricity: without Medupi, "the consequences for our subregion will be dire as a result of the shortages that will ensue."[57] Energy shortfalls, occurring because of the absence of new capacity, "would hinder the economic development of the region. Botswana, Lesotho, Namibia, Swaziland, and Zimbabwe rely on Eskom for their electricity."[58,59] Medupi seems to matter more than ever: "The entire Southern Africa will benefit from the project … Medupi power project will stabilise the electricity trade in the Southern Africa Region."[60]

In addition, the theme of economic development encompasses industrial competitiveness—South Africa's minerals-energy complex—which is historically rooted in cheap energy access.[61,62] While this element of the Medupi project was de-emphasized during the controversial debates over the World Bank loan (collapsed into vaguer references to a more general "economy"), it can nonetheless be clearly discerned in earlier discussions of the plant. As the African Development Bank explained:

> Infrastructure bottlenecks pose significant constraints to private investment in [the] mining and manufacturing sector. Early in the energy crisis in January-February 2008, a near collapse of the grid forced mines to close for

days, and losses from gold and platinum alone, which account for 25% of South Africa's exports, were estimated at ZAR200 million (roughly US$14 million) per day during the closure.[63]

Consistent with the globalized status of these industries' commodity markets, the Medupi project is also explicitly tied to discourses of competitiveness:

> The decision to invest in this infrastructure was borne not only out of necessity, following increased power outages as capacity became strained, but also out of the realization that South Africa is playing in an increasingly globalised world, and is competing with some of the best countries in the world to attract investment opportunities,

explained former DPE minister, Alec Erwin. "Infrastructure that is world-class and efficient is also a key if local companies are to have a fair chance at competing with other world-class companies in the global economy."[64]

Significantly, this economic priority was depicted as outweighing all others: "This was not just playing around with little ideas about having nice clean energy—it was about gearing this country for moving forward," explained the DPE.[65] The "ultimate goal," as defined by Erwin, is "creating more jobs and creating a more vibrant economy with equal opportunities."[66] This is why supporters are equally comfortable promoting the project as growth during good economic times—Medupi-generated "security of supply" was marked in 2007 as "a key priority, especially in light of the significant economic growth that the country has experienced"[67]—as they are promoting it as a solution to crisis during worse times. Rapidly deployed coal-fired power sits at the intersection of both priorities:

> … [A]s the economy recovers from the global crisis, without additional generation capacity, electricity supply will become a "binding constraint" to growth and job creation. In other words, the economy will plateau and additional jobs will not be created: a dangerous scenario for a country with high unemployment and high instances of social disquiet arising from pervasive exclusion.[68]

In this fear of social unrest, the greatest economic threat of energy supply constraints—and perhaps the largest underlying justification for the project—becomes clear: the interests of the ruling party depend on immediately placating the economic worries of its constituents, while the interests of foreign lenders lie in widening and strengthening the productivity of existing markets.

Environmental sustainability

While the imperative of economic growth drove the initial decision to implement Medupi, more recent rhetoric has framed it as an environmentally sustainable enterprise in line with South Africa's global climate leadership. This discourse appears to have evolved over time, as the GoSA formulated more specific reasons for the plant's environmental soundness and began to pitch the plan as an *opportunity* rather than a global warming disaster. Virtually all documents supporting Medupi from 2009 onward extensively discuss its net-positive implications for a transition to clean energy—a drastic shift, considering that the 458-page Environmental Impact Assessment spends less than a single page discussing the risk of GHG emissions.[69] Internal, declassified World Bank documents explicitly identify the "high" risk of "negative public perception" from its financing of climate change, and vaguely reference the creation of a "robust communication strategy" to avoid this negative perception.[70] Further, "The Bank ... will help GOSA implement a viable communications strategy to explain steps taken to address climate change and energy needs."[71] As such, this strain of discourse underscoring "environmental sustainability" ought to be understood as manufactured, strategic, and supplementary to underlying economic priorities.

The core of this sustainability discourse rests on the GoSA's long-term National Climate Change Response Strategy, situating Medupi within a larger agenda for low-carbon development While this appeal to long-term planning contrasts with the short-term economic need for greater energy accessibility, it is articulated with equal measure. As the World Bank declares, "In making the decision, an overwhelming majority of the Executive Directors voted in favor of the loan with only three abstentions. South Africa was applauded on its commitment to the climate change agenda and the pace set on renewable energy."[72] Hogan likewise frames the Medupi loan as emblematic of "our unambiguous commitment to the introduction of cleaner energy technology."[73] And the Department of Energy contends that in the context of Medupi, "South Africa takes its environmental responsibilities seriously and with the fortitude that it surely deserves."[74]

More specifically, the argument for sustainability centers around three interlinking contentions—Medupi's supercritical components, the renewable elements of the World Bank loan, and GoSA's climate plans—which are tightly interwoven to magnify the perception of its environmental friendliness. "The project as conceptualized is technically sound, socially acceptable and environmentally sustainable," writes the AfDB. "The Medupi power plant employs super-critical boiler and dry cooling technologies which contribute to sustainable economy, not harmful to the environment. Eskom is committed to climate change mitigation initiatives."[75] The World Bank similarly links these criteria, explaining that:

> the Medupi power plant is the first in Africa to use the cleaner coal 'supercritical' technology, the same technology used in most high-income countries for new coal power generation ... The railway line associated with this loan will substantially reduce the greenhouse gas emissions from moving coal in trucks.[76]

In this way, proponents frame the economy as a short-term priority while the environment is relegated to long-term "planning for our future energy needs."[77] The distinction—at least conceptually—resolves the tension between the two aims. The head of the World Bank's Africa Energy Group, Subramaniam Vijay Iyer, underscores the ambiguity:

> this is very challenging project, and it strikes a balance between the imperatives that South Africa faces in order to ensure a supply of energy to the people of South Africa, while also charging a part toward a greener energy future.[78]

Characterizing the response to climate catastrophe as not simply future-oriented but future-dependent (deferring a transition to clean energy until a later date) is part and parcel of the World Bank's carefully crafted communications strategy: "The main messages are that in the short-term coal investments will ensure energy security, thus boosting the economy, while investment in renewables will help South Africa meet its low carbon strategy and COP15 commitments."[79]

Energy security

A final discursive strand signifies the importance of Medupi's contribution to energy availability and access. References to maintaining the "security" of South Africa's energy supply are pervasive in the literature surveyed, and this ought not be too surprising. If economic growth is the stated political goal of Eskom's expansion, energy security is the means by which it occurs. "The premise upon which the World Bank loan application for Eskom was made," argues the DEA, is "based on the fundamental belief that developing countries must be allowed to develop their energy security for their populations, in the most cost effective and sustainable manner."[80] In supporting Medupi, the DoE also writes of the "re-emergence of energy security as a major policy imperative,"[81] and Jacob Maroga, Eskom's chief executive, affirms that security of supply "remains Eskom's fundamental responsibility."[82]

Energy security also provides a specific mechanism for poverty alleviation, a compelling argument against the human costs of inaction. The World Bank states it quite simply: "We believe that there can be no poverty reduction without power."[83] "Without reliable energy, the basic services that people in rich countries take for granted cannot be offered ... [f]ailing to address South Africa's energy crisis will have dire consequences for the poor."[84] Medupi "will play an important part in promoting 100% access to electricity" and "aims to benefit the poor directly"[85] and "it will contribute to 'poverty reduction'."[86] The DPE and Eskom also connect the Medupi plant with poverty reduction and the GoSA's commitment to universal access by 2014. It is likely that the allure of "energy security" lies in its discursive potential to unite a deeply divided nation over the issue—"the project will feed into national electricity grid and the entire country will reap benefits through reliable power supply."[87]

Inevitability

The themes of economic development, environmental sustainability, and energy security are given fixed, final and uncontestable force by continual allusions to inevitability, which (1) posit the consequences of not building Medupi as so dire as to be unacceptable and unimaginable, and (2) recognize that the ongoing nature of the procurement and construction process already closes the space for a decision.

The first contends Medupi is inevitable because "building no domestic generation capacity is economically and politically unsustainable and thus, it is rejected."[88] Renewable energy options remain too expensive—estimated close to $20 billion[89]—meaning that a renewable transition is not "feasible" in the near-term.[90] Inevitability rhetoric has operated through the entire build process of the project thus far. The first environmental impact assessment notes, "the no-go option is not considered as a feasible option on this proposed project."[91] More recent language remains unchanged: "We therefore want to state quite categorically that the completion of the Medupi Power Station is inevitable and the acquisition of the World Bank loan should therefore be supported."[92] This belief in inevitability is born out of an understanding of the consequences of energy supply shortfalls as unthinkable. "Never before has Eskom had a single project underway on which so much and so many is dependent," explains Eskom's managing director Brian Dames. "The absolute importance of bringing this project to successful completion and on time just cannot be emphasized enough."[93] "We do not have the luxury" to find another revenue stream, lamented the DoE.[94] Here, as Escobar predicts,[95] pro-Medupi institutional discourse paints Medupi's significance as incalculable, exceeding explanation and comprehensible only by way of expert knowledge from "development professionals."

The second element of inevitability is rooted in the material reality of the plant's ongoing construction. Procurement and site preparation began in early 2006, which means the initial phases of the project were not subject to public debate or consultation. Thus none of the reasons that the GoSA uses to justify the project supersede its primary and initial motivation to increase electricity supply to keep pace with economic growth. The GoSA has always known the trajectory of Medupi, which means the outcome of the debate over the World Bank loan four years into the project is predetermined—failing to receive the loan would simply delay the inevitable.

Discursive coalitions against the Medupi clean coal facility

I found three particularly salient themes at the heart of the campaign against Medupi: (1) a theme of maldevelopment and secrecy that characterizes the project as a collusion between the corrupt ANC, energy-intensive industry, and the imperialist World Bank; (2) a theme emphasizing both Medupi's local and global environmental degradation; and (3) a theme of energy poverty, which highlights the disproportionate impact of energy price increases within the legacy of apartheid and concludes with the imperative of a non-fossil-fuel energy future.

Maldevelopment

The first theme, tentatively labeled "maldevelopment," a phrase drawn from Samir Amin,[96] indicts Medupi as capitalist profiteering designed to enrich greedy ANC bureaucrats, industrial corporations, and the World Bank at the expense of the South African people. In this view, Medupi's construction and eventual production will line the pockets of a few enterprising industrial, financial and political elites while sustaining an exploitative, export-oriented economic system at the expense of the vast majority of (black) South Africans. As Africa Action's Michael Stulman explains: "This is one of those stereotypical development disaster stories."[97] Consistently appearing throughout the anti-Medupi discourse, this argument calls into question the intentions of the institutions behind Eskom's build plan, and thus complicates not simply the project's concrete impact but the discourses that support it. The contention also hinges on an independent claim of secrecy, both in the distribution of Medupi's electricity, through opaque, "apartheid-era contracts" that guarantee cheap energy to industries like smelting or mining, and in the contracting process behind the construction of the plant itself, from which the ANC stands to profit through its in-house investment firm's 25 percent share in Hitachi Power Africa.

The Medupi plant, in the view of the civil society organizations I studied, will benefit only "energy-greedy multinational corporations" and not the "people" of South Africa.[98] Medupi, as articulated in the words of Earthlife Africa Jhb's Tristen Taylor, will "hurt our economic and social development."[99] This allegation of *maldevelopment* builds on the same interpretation of South Africa's minerals-energy complex as the pro-Medupi camp—energy as the lifeblood of South African development and economic competitiveness—but calls into question the type of growth the system creates. Neither side disputes the fact that "[c]heap power has remained at the core of industrial policy ... irrespective of who holds political power"[100]; the debate is rather over who benefits from the system. This move enables an inversion of the economic crisis discourse that the GoSA used to justify the World Bank loan, where "infrastructure investments are now proclaimed as a 'countercyclical preparation' for the next boom."[101] As Earthlife's Taylor comments:

> with massive disconnections looming due to a doubling of electricity tariffs, a million jobs lost last year, and an effective 40% unemployment rate, one would think that poverty eradication would be foremost in the World Bank and the South African Government's mind. None of Medupi's output will be for the poor but will be used to service multinational firms.[102]

When taken to the extreme, this discourse paints Medupi as an explicit link between an apartheid development narrative that privileges South African corporations and the economic history of Western imperialism. For the anti-Medupi coalition, it is "foreign corporations" that "benefit from the world's cheapest electricity," that is, a price differential that only Medupi can sustain.[103] These:

> industries in turn are mostly geared for export in line with the World Bank's promotion of export oriented production. The goods are then consumed primarily in developed countries. Further, many industries are established with foreign direct investments .... because their headquarters are in London, Melbourne, New York, Toronto, Zurich and other off-shore sites, a substantial portion of profits is returned to rich countries, exacerbating the poor countries' balance of payments deficit.[104]

Medupi ensures a "further entrenching of an export-orientated economy in raw materials," an "economic model that has consistently failed, for the last hundred years, to eradicate poverty in the country."[105]

Crucially, the theme of *maldevelopment* is inseparable from the allegations of corruption and anti-democratic practice that saturate discourse from all the anti-Medupi NGOs, as both speak to the intentions of the project's financiers rather than its consequences. However, maldevelopment also indicates a difference within civil society opponents—critiques by larger, international environmental organizations like the WRI and WWF are process-oriented, emphasizing undemocratic decision-making structures, while South African environmental justice groups directly indict the morals of the individual participating decision-makers. "Electricity planning processes to date ... have been neither transparent nor inclusive," the WRI writes in its only public document relating to Medupi.[106] The WWF also voices concerns with secrecy, writing that "[t]he 'Integrated Energy Plan' justifying Eskom's build plan was gazetted in a 'midnight regulation'" which "ignores the requirements of the Energy Act of 2008 for transparent public consultation."[107] In contrast, groundWork stands up against the "managers of SA Inc" and Eskom's "bureaucratic fat-cats."[108] In both cases, the nongovernmental organization is the reasonable, democratic truth-seeker: aiming to reveal what "the truth is" and uncover how "in reality" the Medupi project will work. This is a crucial point to which I (and the NGOs) will return.

Environmental degradation

While the critique of Medupi is generally situated within a narrative of elitist, corporate-driven development, none of the anti-Medupi documents surveyed fail to mention the project's environmental impact. This pervasive theme of *environmental degradation* condemns the coal-fired Medupi plant as a substantial contributor to global climate change and local environmental degradation, and decries

GoSA and its financiers as hypocritical and short-sighted. Support for the project is declared "an assault on the livelihoods and way of life of global citizenry ... akin to fighting a fire with petrol,"[109] as "disastrous for the environment,"[110] and showing "no regard for the state of the world's climate and environment."[111] Figure 5.2 shows one protest in Johannesburg attacking clean coal for "killing" and "choking" South Africa with pollution. Medupi is described as "one of the largest and dirtiest coal-fired power plants in the world"[112] and considered antithetical to sustainable development. The "flies in the face of the World Bank's claims of ... sustainable development,"[113] and the World Bank's loan as support for "large-scale unsustainable power generation in an age when such projects should not even be considered."[114]

In particular, Medupi's opponents position its climate impact as extending beyond actual emissions into its precedent for broader South African climate policy. WRI, for instance, worries that Medupi "may have problematic implications for environmentally and socially sustainable development in South Africa."[115] Similarly, the petition signed by over 300 NGOs evokes South Africa's disproportionately large contribution to the continent's emissions—what it calls Eskom's "special responsibility to Africa"—to critique Medupi as sinking South Africa into a "much deeper 'Climate Debt' to Africa," concluding that "African civil society unites with SA critics of Eskom's irresponsible climate-denialist projects."[116] References to South Africa's "climate debt" and the environmental responsibility it implies appear throughout anti-Medupi discourse, reinforcing the cost of Medupi not simply as a contributor to climate change but as a symbol of GoSA's failure to move toward broader environmental sustainability. "At the heart of the matter," as Earthlife's Taylor describes, "is the energy future of the entire planet."[117] While characteristically hyperbolic, Taylor's move to place Medupi in the wider debate over the future of sustainable development illustrates both a prevalent metonymic tactic and a compelling reason why Medupi provokes a backlash from so many environmental groups outside South Africa.

Depicting Medupi as environmentally hypocritical also involves specific attacks on South Africa's long-term climate mitigation plans and the World Bank loan's modest clean energy provisions. GoSA's climate commitment is chalked up as mere posturing: "It seems like South Africa just want to shine in the negotiations and then continue with business as usual," Ferrial Adams of Earthlife explains. "They know very well that their promise of emission reductions is not legally binding."[118] Medupi exists, "irrespective of any rhetorical devotion to climate mitigation,"[119] which reveals GoSA's plan as "neither robust nor ambitious;"[120] that is, as insincere and misleading. Likewise, groundWork decries the loan's various "low-carbon" provisions as a "renewable energy fig leaf."[121] Emphasis on Medupi's status as a supercritical plant is "green spin,"[122] and claims of carbon savings from the new railroad constitute a "poor greenwashing attempt."[123] It is salient to recognize that GoSA has not yet issued a legally binding emissions plan. Pro-Medupi descriptions of Medupi's place in South Africa's wider long-term,

FIGURE 5.2 Protesters from Earthlife Africa in Johannesburg, South Africa, May 2015
Source: Author

low-emissions trajectory reference either the National Climate Change Response Strategy, which is non-binding or the "Long-Term Mitigation Scenarios" (LTMS), which is not a legislative agenda but rather an exploration of possible emissions reductions strategies.

The second part of Medupi's purported unsustainability is its impact on the local environment and community of Limpopo, understood as a series of incalculable—and therefore unacceptable—externalities. These are explained as "hidden costs," including "health impacts from air pollution, elevated SO2 levels, and mercury residues in their water, air and land; [and] constrained access to water."[124] Anti-Medupi environmental organizations also describe the destructive effects of the acid drainage accompanying the increase in coal mining necessary to fuel the Medupi plant. The cumulative impact of this local pollution is described qualitatively, and used to call into question the dominant, economic calculus that justified the GoSA's decision. Local impacts, then, raise:

> very important questions about the real costs and benefits of this project. The Bank overemphasis benefits in terms of 'poverty alleviation', 'energy security' and 'economic growth', whereas deliberately ignoring social and environmental costs associated with coal-fired power plants and associated coal mines.[125]

Limpopo's depiction as a "sacrifice zone" is an ambiguous but persuasive rhetorical abstraction. Its outrage is in line with the NGO joint statement, which writes that these environmental consequences outweigh even gains in income calculated by "conventional measures."[126] This highlights a defining element of the theme of environmental degradation, which involves an implicit appeal to a prioritization of the environment as an inherent, absolute good.

The trope of local unsustainability also includes an appeal to the sanctity of the "local community," a counter-narrative that attempts to trump the elite-centered Medupi front. To the South Durban Community Environmental Alliance (interestingly located more than 900 miles from Lephalale), Medupi threatens "hundreds of people that live in those communities," who "will be forced out of them and lose their cultural, their agricultural land."[127] The project poses "a grave threat to communities" for groundWork as well.[128] Here, anti-Medupi discourse attempts to speak for the people of South Africa: the battle is not between environmentalists and economists but rather between the authentic, local community and the impersonal, corporate state. Who these communities actually *are*, however, is just as unclear as the identity of the local groups the GoSA claimed to consult. The complaint from Lephalale residents was "submitted by Earthlife Africa and ground-Work on behalf of affected community members,"[129] who remain unnamed. Likewise, the only example amidst all of the pro-Medupi documents of an affected resident is "Caroline Ntapoane," who "hails from South Africa's polluted industrial heartland near Sasolburg" and "knows first-hand what communities have to look forward to."[130] However, the document fails to mention that Sasolburg is located 350 km from Lephalale, or that Ntapoane is referenced only once elsewhere, as a representative of the Vaal Environmental Justice Alliance.[131]

Finally, the discourse of environmental degradation includes a temporal appeal to futurity and farsightedness that reverses the similar appeal made by GoSA. Immediately apparent is the oft-repeated concept of lock-in, which recognizes that the multi-decade lifespan of Medupi will extend its climate impact far into the future. The World Bank is condemned for its short-sightedness: the "major economic actors have focused on who pays but not on what they are paying for."[132] "Today's children will judge them harshly in the years to come," for they "do not have South Africa's long-term interests at heart."[133] This emphasis on far-sightedness is used to bolster the case for action today: "The Earth can no longer wait, we have to stop this environmental injustice now!"[134] The WRI similarly argues that debate is "especially crucial" for "such long-term investments."[135] The environmental degradation contention therefore includes not only accusations of hypocrisy but appeals to particular valuations of the environment and future generations.

Energy poverty and racism

I identify a separate theme of *energy poverty* as a deliberate counter-discourse against that of energy security. Energy poverty contends that electricity rate hikes used to finance Medupi will endanger the livelihoods of already-impoverished and energy-starved South Africans. It also claims that Medupi impedes the clean-energy economy, which is envisioned as an alternative mode of creating opportunity and universal energy access. References to the coal-fired plant's detrimental impact on access are nearly universal within the discourse studied. The loan "flies in the face of the World Bank's claims of alleviating poverty;"[136] "will not help the poor;"[137] "would actually decrease access to electricity for poor households"[138] and "will not alleviate 'energy poverty' in South Africa but rather aggravate poverty [and] worsen ongoing inequities in access to electricity" by "any calculation;"[139] and "entrenches suffering by imposing 'cost recovery' on people who cannot afford it."[140] Specifically, NERSA's legislated electricity tariff increases mean a "typical township household will face a 2009 to 2012 monthly price rise from R360 ($48) to R1000 ($130)."[141]

The focus on energy poverty, like the alleged government-industry pacts that exemplify maldevelopment, is inevitably placed in the context of South Africa's history of apartheid. South Africa is, after all, the "most unequal large society in the world,"[142] where electricity cables, as with other lines of power, are drawn largely along racial divides. Medupi's electrical output will therefore mirror "apartheid-era Eskom's distribution of power to whites."[143] Maldevelopment converges with concrete measures like rate hikes and cost-recovery to exacerbate energy poverty: "[i]n sum, the World Bank is repeating one of the world's most tragic episodes: apartheid empowerment of corporations and whites, and impoverishment of black South Africans."[144] Anti-Medupi groups, recognizing GoSA and the World Bank's repeated references to energy insecurity, also highlight what they consider an absence of explanation detailing precisely *how* Medupi will help poor South Africans that lack electricity. For instance, the WWF notes that the World Bank failed "to provide specifics of how the claimed public benefits, including contribution towards achieving universal access to energy services, will be achieved."[145]

This discourse of energy poverty coexists with the clean energy economy, imagined as simultaneously inclusive and sustainable. Certainly, renewable energy is also relevant to the project's environmental consequences, but I found the discourse more interesting in its explicitly articulated relation to eradicating energy poverty:

> We see renewable energy, not coal-fired power stations, as the optimal development path for Southern economies, creating more jobs, building local manufacturing capacity, and avoiding the environmental mistakes of Northern countries.[146]

In addition to the joint NGO statement, Greenpeace urges large-scale renewables development "to help facilitate green development pathways" and "create green jobs."[147] Elsewhere, opposition voices claim that renewables "cost less in long-term savings from reduced energy volatility and job gains"[148] and will create "a new, job-intensive industrial base."[149] The devastation wreaked by Medupi on the poor is thus identified not simply in immediate rate increases for South Africans but also with the equitable energy transition that it precludes.

Democracy

In line with the predictions made by Hajer that distinct narratives cohere into singular discourses, themes of maldevelopment, environmental degradation and energy poverty ultimately coalesce around a deeper claim and final theme of *democracy*, which posits the anti-Medupi campaign as a global struggle for transparency and fact-based policy. The agents resisting Medupi claim to represent a "global campaign," one "endorsed by hundreds of organizations that represent millions of concerned citizens, community, environmental, labor and academic constituencies, in South Africa, the rest of the African continent and the world at large."[150,151]

It is, in other words, a hodge-podge of various civil society organizations—a "broad based coalition" comprising "local civil society groups" and "global NGOs"[152] including not just "social movements, environmental groups, academic institutions and trade unions" but "[p]eople from all over the world."[153] While pro-Medupi actors buttress their case by appealing to their bureaucratic expertise, the opposition attempts to generalize its institutional location, and to speak for what it considers "universal condemnation of Eskom."[154] This is particularly noteworthy given the incredibly diverse interests represented by the each of the civil society groups (environmental justice, human rights, and conservationism), which additionally complicate their claim to present an apparent unified front.

This presumption of universality emerges out of a wholly constructed appeal to impartiality, with the implication that fair debate could not conscionably support the project. The "underlying problem" of Medupi is simply "that there has been little transparency or public debate of the assumptions that underpin long-term plans."[155] Similarly, Eskom and the World Bank are simply described as exhibiting "an almost complete lack of effort" to "evaluate clean energy alternatives"[156] and as failing to "adequately recognize" the potential of renewable energy.[157] Poor people are "ignored," while environmental externalities are not incorporated in the "project analysis."[158] Anti-Medupi groups therefore position themselves in opposition to what they perceive as indifferent self-interest, advocating "a genuinely democratic and developmental strategy;"[159] "open, fact based debate on all available energy options;"[160] "a transparent consensus-reaching approach to future energy planning in the country;"[161] and the simple request that Congress "demand accountability."[162]

Just as the concept of inevitability structures the pro-Medupi discourse, this affirmation of objectivity and transparency propels the righteousness of the cause and the movement of the discourse itself. Hence, after Medupi, it already begins to repeat itself:

The approval of this dirty loan marks not an end, but a beginning. The World Bank has already said it would loan South Africa another $1 billion, rubbing salt into the wounds of South Africans. This is the beginning of the resistance, both in South Africa and across the globe, against future World Bank and coal shenanigans. A new energy future is possible, and, indeed, the only choice we have.[163]

Conclusion

My discursive investigation of the debate over Medupi reveals a set of tightly interdependent narratives summarized in Table 5.2. Both the coal plant's proponents and opponents actively and strategically construct their discursive framing of the project, shifting their positions over time and persistently signifying and resignifying Medupi for their own ends. Ironically, the very "people" and "communities" the plant is accused of empowering or exterminating become mere abstractions to be strategically invoked "for" or "against" the project. Each side betrays an underlying ideological frame that structures their discourse as an absolute, indisputable conclusion. The set of institutional actors behind Medupi—government agencies such as Eskom, DPE, and DoE, and multilateral lending banks such as the World Bank and the AfDB—propagate a discourse of inevitability that considers the imperatives of economic growth and energy security as incontrovertible justifications for the project. In much the same way, the most vocal actors opposing the project—civil society organizations like groundWork, Friends of the Earth South Africa, Earthlife Africa Jhb, Africa Action, Greenpeace, and the WRI—generally articulate a discourse posited as democratic, neutral, and universal which claims that Medupi is an unquestionably corrupt, environmentally destructive endeavor designed to enrich a few corporate and political fat cats at the expense of collective energy security.

TABLE 5.2 Summary of discursive themes and institutions

	Institutions	Theme 1	Theme 2	Theme 3	Convergence
Pro-Medupi	World Bank, AfDB, DPE, DoE, DEAT, Eskom, GoSA	Economic development	Environmental sustainability	Energy security	Inevitability
Anti-Medupi	Africa Action, Earthlife Africa Jhb, Greenpeace, groundWork, South Durban Community Environmental Alliance, WRI, WWF, WESSA	Maldevelopment	Environmental degradation	Energy poverty	Democracy

Source: Author, based on Rafey, W., and B. K. Sovacool 2011, August[164]

These two convergent discourses offer insight for both scholars and policy-makers. First, the proliferation of contested narratives around Medupi—that the project can be situated simultaneously as energy security and energy poverty, as communitarian and authoritarian, as environmentally destructive and environmentally sustainable—underscores the importance of locating discourses within concrete institutions. Discourses constitute specific, strategic crystallizations of power that are both changeable and self-reinforcing. In particular, the concepts of "energy security" and "climate change" cannot be interpreted without reference to their specific locations in people and organizations. Academic work examining particular truth-regimes needs to look not simply at what is being said but *where* it is said and *by whom*. Trust or legitimation of the source is almost as important as the substance or content of the message.

Second, and perhaps most importantly, the sheer intractability of the Medupi debate underscores the need for intervention—by policymakers, analysts, and even activists—to change the terms of the debate itself. Medupi is neither the complete and only solution to energy security, nor is it the sole cause of energy poverty. It certainly does not indicate strong climate leadership, but neither will it entirely destroy the South African natural environment. Its construction need not have been inevitable, but a "democratic truth-seeker" might not have had any other realistic options. The Republic of South Africa's choice to build Medupi reflects a set of decisional constraints, rooted in an assumption of the relative cost of alternative energy. The choice is magnified by the underfunding of Eskom, political pressure to keep prices low due to the impact of price hikes on low-income households, and abundant coal reserves which keep conventional energy relatively cheap.[165] Possible solutions abound: legal-regulatory fixes to place an international price on carbon-intensive commodities,[166] direct transfers of technology to developing countries,[167] integration into low-carbon technology supply chains and global production networks,[168] and accelerated technological innovation[169] are just a few.

As Medupi shows all too clearly, however, building energy infrastructure in the developing world continues to be a bitterly contested battleground. Claiming, for example, that there is "a price to be paid" for Africans' "inaction and lackadaisical attitude" toward climate change[170] is utterly incomplete. Successful climate politics will require more than charged rhetoric or persuasion. Most everyone, after all, wants rain that soaks parched lands and brings prosperity. However, that possibility will depend not on who performs the most compelling rain dance but who will build the most sustainable irrigation system.

Notes

1 Eskom. Medupi Power Station Fact Sheet. November. www.eskom.co.za/content/NB%200002MedupiPowerStation.doc.
2 PB (Parsons Brinckerhoff). 2010. Medupi Power Station. www.pbworld.com/projects/featured/medupi_power_station_26825.asp.
3 Eskom. 2015. Kusile and Medupi Coal-Fired Power Stations under Construction. COP 17 Fact Sheet.

4 AFDB. 2009b. 'Africa Needs to Significantly Scale up Investments in Both Renewable and Low-Emission Fossil-Fuel Technologies,' Says AfDB Infrastructure Director, Gilbert Mbesherubusa. 26 November. www.afdb.org/en/news-events/article/africa-needs-to-significantly-scale-up-investments-in-both-renewable-and-low-emission-fossil-fuel-technologies-says-afdb-infrastructure-director-gilbert-mbesherubusa-5423/.

5 AFDB. Project Appraisal Report: Medupi Power Plant, South Africa. 14 October. www.afdb.org/fileadmin/uploads/afdb/Documents/Project-and-Operations/South%20Africa%20-%20Medupi%20%20Power%20Project.pdf

6 Earthlife Africa Jhb. 2010b, March 25. Eskom's World Bank Loan–A Climate Shame! Press Release. www.earthlife.org.za/?p=902.

7 Njobeni, S. 2010b. South Africa: World Bank Defends Loan to Eskom. Business Day. 8 March. http://allafrica.com/stories/201003081445.html.

8 Rafey, W., and B. K. Sovacool. 2011, August. Competing Discourses of Energy Development: The Implications of the Medupi Coal-Fired Power Plant in South Africa. Global Environmental Change 21 (3): 1141–1151.

9 Pegels, A. 2010, September. Renewable Energy in South Africa: Potentials, Barriers and Options for Support. Energy Policy 38 (9): 4945–4954.

10 Sovacool, B. K., and W. Rafey. 2011, January/February. Snakes in the Grass: The Energy Security Implications of Medupi. Electricity Journal 24 (1): 92–100.

11 Baker, L., and B. K. Sovacool. 2017, September. The Political Economy of Technological Capabilities and Global Production Networks in South Africa's Wind and Solar Photovoltaic (PV) Industries. Political Geography 60: 1–12.

12 Freund, B. 2010. The Significance of the Minerals-Energy Complex in the Light of South African Economic Historiography. Transformation: Critical Perspectives on Southern Africa, Number 71: 3–25.

13 Ferrey, S. 2009, April. The Missing International Link for Carbon Control. The Electricity Journal 22 (3): 17–28.

14 Sovacool, B. K., C. Cooper, and P. Parenteau. 2011, August. From a Hard Place to a Rock: Questioning the Energy Security of a Coal-Based Economy. Energy Policy 39 (8): 4664–4670.

15 Sovacool, B. K., M. A. Brown, and S. V. Valentine. 2016. Fact and Fiction in Global Energy Policy: Fifteen Contentious Questions (Baltimore: Johns Hopkins University Press).

16 World Coal Mining. 2013. Industry Overview. www.mbendi.com/indy/ming/coal/p0005.htm.

17 IEA, World Energy Outlook (2011).

18 International Energy Agency. 2014. World Energy Outlook 2014 (Paris: OECD).

19 IEA. Coal's Share of Global Energy Mix to Continue Rising, with Coal Closing in on Oil as World's Top Energy Source by 2017, December 17, 2012, available at www.iea.org/newsroomandevents/pressreleases/2012/december/name,34441,en.html

20 Yang, A., and Y. Cui. 2012, November. Global Coal Risk Assessment: Data Analysis and Market Research (Washington, DC: World Resources Institute).

21 International Energy Agency. 2008. World Energy Outlook 2008. www.worldenergyoutlook.org/docs/weo2008/WEO2008_es_english.pdfright-pointing angle bracket

22 To Clarify, in This Chapter We Understand "Clean Coal" to Refer to a Collection of Four Technologies and Processes: (A) Supercritical Pulverized Coal Plants that Boost Thermal Efficiency by Operating at Higher Temperatures, (B) Integrated Gasification Combined Cycle (IGCC) Plants that Use Chemical Processes to Gasify Coal and Remove Sulfur and Mercury, (C) Pressurized Fluid Bed Combustion Plants that Use Elevated Pressure to Capture Sulfur Dioxide and Nitrogen Oxides, and (D) Carbon Capture and Storage (CCS) Techniques Such as Deep Underground Geologic Formations that are Engineered to Capture and Store Excess CO_2.

23 U.S. EPA, Clean Air Markets. 2012, July 25. Section 415 Clean Coal Technology Regulatory Incentives. www.epa.gov/airmarkets/progsregs/arp/sec415.html.

24 Colinet, J. et al. 2010. Best Practices for Dust Control in Coal Mining (Pittsburgh, PA, and Spokane, WA: National Institute for Occupational Safety and Health, Office of Mine Safety and Health Research).

25 Özgen Karacan, C. et al. 2011, 1 May. Coal Mine Methane: A Review of Capture and Utilization Practices with Benefits to Mining Safety and to Greenhouse Gas Reduction. International Journal of Coal Geology 86 (2–3): 121–156.

26 Franks, D. M., D. Brereton, and C. J. Moran. 2010. Managing the Cumulative Impacts of Coal Mining on Regional Communities and Environments in Australia. Impact Assessment and Project Appraisal 28 (4): 299–312.

27 Carbon Sequestration Technology Roadmap and Program Plan 2006, DoE, 2006: 11–12. Available at http://fossil.energy.gov/sequestration/publications/program plans/2006/2006_sequestration_roadmap.pdf

28 Global Carbon Capture and Storage Institute. 2012, October. The Global Status of CCS: 2012 (Melbourne, Australia: Global CCS Institute).

29 Eskom. 2015. Kusile and Medupi Coal-Fired Power Stations under Construction. COP 17 Fact Sheet.

30 Foucault, M. 1979. Discipline & Punish: The Birth of the Prison (New York: Vintage Books).

31 Bakhtin, M. M. 1984. Problems of Dostoevsky's Poetics, edited and translated by Emerson, C. (Minneapolis: University of Minnesota Press), p. 183.

32 Hajer, M. A. 1993. Discourse Coalitions and the Institutionalization of Practice: The Case of Acid Rain in Britain. In Fischer, F. and Forester, J. (Eds): The Argumentative Turn in Policy Analysis and Planning (Durham, NC: Duke University Press), pp. 45–46.

33 Escobar, A. 1995. Encountering Development. The Making and Unmaking of the Third World (Princeton, NJ: Princeton University Press).

34 Doulton, H., and K. Brown. 2009. Ten Years to Prevent Catastrophe? Discourses of Climate Change and International Development in the UK Press. Global Environmental Change 19: 191–202.

35 Sovacool, B. K., and S. Halfon. 2007, April. Reconstructing Iraq: Merging Discourses of Security and Development. Review of International Studies 33 (2): 223–243.

36 Bakhtin (1984: 183).

37 Steinberg, M. W. 1998. Tilting the Frame: Considerations on Collective Action Framing from a Discursive Turn. Theory and Society 27: 845–872.

38 Smith, S. 2002. Historical Sociology and International Relations Theory. In Hobden, S. and Hobson, J. M. (Eds): Historical Sociology of International Relations (Cambridge: Cambridge University Press), p. 241.

39 Litfin, K. T. 1994. Ozone Discourses: Science and Politics in Global Environmental Cooperation (New York: Columbia University Press).

40 Escobar, A. 1984. Discourse and Power in Development: Michel Foucault and the Relevance of His Work to the Third World. Alternatives 10 (3): 377–400.

41 Escobar (1995: 54)

42 Hajer (1993: 44).

43 Escobar (1984: 393)

44 Adams, I. S., and T. O. McShane. 1996. The Myth of Wild Africa: Conservation without Illusion (Los Angeles, CA: University of California Press).

45 Hajer (1993).

46 Escobar (1995).

47 Ferguson, J. 1994. The Anti-Politics Machine: "Development," Depoliticization and Bureaucratic Power in Lesotho. (Cambridge: Cambridge University Press).

48 Bialasiewicz, L. D. et al. 2007, May. Performing Security: The Imaginative Geographies of Current US Strategy. Political Geography 26 (4): 405–422.

49 Rafey and Sovacool (2011: 1141–1151).

50 Parliamentary Monitoring Group. 2010, 12 March. World Bank Loan for Eskom: Media Briefing by Ministers of Energy and Public Enterprises. Parliamentary Monitoring Group. www.pmg.org.za/briefing/20100312-minister-dipuo-peters-statement-world-bank-loan.

51 AfDB (2009c: 13)

52 Eskom. It's All Systems Go for the Start of Boiler Construction at Medupi Power Station. Joint media release from Eskom and Hitachi Power Africa. 16 January. www.eskom.co.za/live/monster.php?URL=%2Fcontent%2FEskom-Hitachi+Joint+Statement+16012010.doc&Src=Item+13938.

53 World Bank. Project Appraisal Document on a Proposed Loan in the Amount of US$3,750 Million to Eskom Holdings Limited Guaranteed by Republic of South Africa for an Eskom Investment Support Project. Report No: 53425-ZA. 19 March. www.wds.worldbank.org/external/default/WDSContentServer/WDSP/IB/2010/04/12/000112742_20100412110336/Rendered/PDF/534250R20101005914.pdf.

54 DPE. South African Government's Response to Questions on the Eskom Loan Application to the World Bank. 8 April. www.environment.gov.za/docs/DocumentHomepage.aspx?type=D&id=3988.

55 World Bank (2010d: 22, 27).

56 DoE 2010c. Comments by Ms Dipuo Peters, MP, Minister of Energy regarding the World Bank loan, Union Buildings. 12 March. www.info.gov.za/speeches/2010/10031214451002.htm.

57 DoE (2010c).

58 Eskom 2010. Eskom aims to turn over new leaf. 27 April. www.eskom.co.za/live/monster.php?URL=%2Fcontent%2FEskom+aims+to+turn+a+new+leaf.doc&Src=Item+13938.

59 World Bank. 2010d. Project Appraisal Document on a Proposed Loan in the Amount of US$3,750 Million to Eskom Holdings Limited Guaranteed by Republic of South Africa for an Eskom Investment Support Project." Report No: 53425-ZA. pp. 4–5. 19 March. http://www-wds.worldbank.org/external/default/WDSContentServer/WDSP/IB/2010/04/12/000112742_20100412110336/Rendered/PDF/534250R20101005914.pdf.

60 AfDB (2009c: 7).

61 Davidson, O., and H. Winkler. 2003, September. South Africa's Energy Future: Visions, Driving Factors and Sustainable Development Indicators. Report for Phase I of the Sustainable Development and Climate Project. Energy & Development Research Centre, University of Cape Town.

62 Fine 2008. Engaging the MEC: or a few of my views on a few things. Transformation: Critical Perspectives on Southern Africa, 71: 26–42.

63 AfDB (2009c: 1).

64 DPE (2007a).

65 Parliamentary Monitoring Group (2010).

66 DPE (2007a).

67 DPE (2007c).

68 World Bank (2010d: 2).

69 Bohlweki Environmental (Pty) Ltd. 2006. South Africa - Eskom Power Investment Support Project: Environmental Assessment (Vol. 6 Of 14): Report for the Proposed Establishment of a New Coal-Fired Power Station in the Lephalale Area, Limpopo Province. Report No. E2279. 22 May. www.wds.worldbank.org/external/default/WDSContentServer/WDSP/IB/2009/11/09/000020953_20091109115541/Rendered/PDF/E22790VOL1060BOX0342035B.pdf, also www.eskom.co.za/live/content.php?Item_ID=2343.

70 World Bank (2010d: 42).

71 World Bank (2010e: 17).

72 GoSA (2010a).

73 DPE (2010b).

74 DOE (2010c).

75 AfDB (2009c: 18).

76 World Bank. 2010a. Eskom Investment Support Project Questions & Answers. p. 5. 14 April. http://siteresources.worldbank.org/INTSOUTHAFRICA/Resources/ Q_A_Eskom_Investment_Support_Project_031810.pdf.

77 GoSA (2010a).

78 Moneyweb. 2010, April 9. World Bank Approves Eskom Loan: Subramaniam Vijay Iyer (Africa Energy Group, World Bank) and Paul O'Flaherty (Financial Director, Eskom). www.moneyweb.co.za/mw/view/mw/en/page295799?oid=479919&sn= 2009+Detail&pid=287226.

79 World Bank (2010d: 42).

80 Dea (2010: 1).

81 DoE. 2010b. Working together to usher in a New Era for Energy Modeling, Forecasting, Planning and Delivery. Budget speech by Minister of Energy Dipuo Peters. p. 3. 20 April. http://www.energy.gov.za/MINISTER_OF_ENERGY_BUD GET_VOTE_SPEECH_2010_FINAL.pdf

82 Eskom (2009b).

83 Sapa (2010).

84 World Bank (2010a: 1).

85 World Bank (2010a).

86 AfDB (2009c: 13).

87 AfDB (2009c: 7).

88 AfDB (2009c: 4).

89 World Bank (2010d: 48–49).

90 World Bank (2010d: 13).

91 Bohlweki Environmental (2006: 10).

92 DoE (2010c).

93 Eskom 2010g.

94 DOE (2010c)

95 Escobar (1995: 52).

96 Amin, S. 1990. Maldevelopment: Anatomy of a Global Failure (Tokyo; London; Atlantic Highlands, NJ: Zed Books).

97 Goldenberg, S. 2010. World Bank's $3.75bn Coal Plant Loan Defies Environment Criticism. The Guardian. 9 April. www.guardian.co.uk/business/2010/apr/09/ world-bank-criticised-over-power-station

98 groundWork. 2010b. World Bank's $3.75 Billion Coal Loan to Eskom: Neither Green nor Just. Authored by Bobby Peek. 8 April. http://www.foe.org/world-banks-375-billion-coal-loan-eskom-neither-green-nor-just

99 Ross, K. 2010, April, 13. Eskom Defends Power Plan. The Pretoria News, p. 5. Also: www.sundayindependent.co.za/index.php?fSectionId=1043&fArticleId= vn20100413043209717C202978

100 groundWork. 2009b. The World Bank and Eskom – Banking on Climate Destruc-tion. Written by David Hallowes. 19. December. http://www.groundwork.org.za /Publications/worldbankeskom09.pdf

101 groundWork (2009b: 20).

102 Africa Action. 2010a, April 6. South Africans Say 'No' to Eskom Coal – Project-Affected Communities Take Their Case to the World Bank Inspection Panel. Press Release. www.africaaction.org/inspection-panel-pr.html

103 groundWork et al. 2010. Annex two: Statement from SA and Africa Civil Society on Eskom's Proposed $3.75 billion World Bank loan. Statement signed by over 300 NGOs, including groundWork, Africa Action, Earthlife Africa Jhb, WWF, and WRI. 2. No specific date cited. http://www.groundwork.org.za/Publications/ EskomFinalDocs/CivilSocietyCritique.pdf

104 Ibid.
105 Earthlife (2010a). Earthlife Africa Jhb. 2010a. World Bank's Climate and Governance Disaster. Press Release. 9 April. http://www.earthlife.org.za/?p=914
106 World Resources Institute. 2010, March 8. The World Bank Eskom Support Program. Authored by Smita Nakhooda, Fellow in the Institutions and Governance Program at the World Resources Institute. www.wri.org/stories/2010/03/world-bank-eskom-support-program
107 WWF. Just Come Clean. The Star. Op-ed written by Peet du Plooy. 22 January. http://secure.financialmail.co.za/10/0122/opinion/bopinion.htm.
108 groundWork. 2010e. South Africans say no to Eskom's R29 billion World Bank loan. p.2. 16 February. http://www.groundwork.org.za/Publications/EskomFinal Docs/EskomWBloanpressstatement16%20February10.pdf.
109 Earthlife (2010a).
110 Friends of the Earth International. 2010. World Bank Considers Billions for Dirty Coal in South Africa. www.foe.org/world-bank-considers-billions-dirty-coal-south-africa
111 groundWork/Friends of the Earth South Africa. 2010a. Eskom loan blackens the World Bank's name. Opinion piece by Bobby Peek, director of groundWork. 16 April. http://www.brettonwoodsproject.org/art-566122
112 Africa Action (2010b).
113 groundWork. 2010h. "Keep the Coal in the Hole!": Critics of World Bank financing for Eskom announce global "No Coal Loan" campaign. Description of press conference proceedings. 22 February. http://www.groundwork.org.za/Press% 20Releases/21Feb10WorldBankEskom.asp
114 van Den Bosch, S. 2009. Africa: South Africa's Empty Promise. Inter Press Service (Johannesburg). http://allafrica.com/stories/printable/200912150890.html
115 Wri (2010).
116 groundWork et al. (2010: 1).
117 Ross (2010).
118 Van Den Bosch (2009).
119 groundWork (2009b: 20).
120 Ibid.
121 groundWork. 2009a. Is the South African government serious about climate change? Ask the World Bank! Press Release. 14 December. http://www.groundwork.org.za /Press%20Releases/14Dec09EskomReport.asp
122 groundWork (2009b: 17–18).
123 groundWork. 2010g. Response to World Bank-Eskom Panel Report and Fact Sheet. p. 4 http://www.groundwork.org.za/Publications/EskomFinalDocs/Responsetothe WorldBankpanelreportandFactSheet.pdf
124 Africa Action (2010).
125 groundWork. 2010c. Re: Proposed 3.75 billion USD loan by World Bank to South African power utility Eskom. Letter to Eli Debevoise, Executive Director of the World Bank, authored by Bobby Peek. p. 3. 1 March. http://www. groundwork.org.za/Publications/EskomFinalDocs/lettertoWB%20ED.pdf
126 groundWork et al. (2010: 2).
127 D'Sa, D. 2010. World Bank Approves Multi-Billion-Dollar Loan for Coal-Fired Power Plant in South Africa. Interview conducted by Democracy Now!. Desmond D'Sa is coordinator of the South Durban Community Environmental Alliance. 8 April. www.democracynow.org/2010/4/9/world_bank_approves_multibillion_dollar_loan
128 groundWork (2010c: 4).
129 Africa Action (2010a).
130 Ibid.
131 Friends of the Earth International (2010).
132 groundWork (2009b).

133 Earthlife (2010a).
134 Earthlife (2010b).
135 Wri (2010).
136 Africa Action (2010b).
137 D'Sa (2010).
138 Friends of the Earth International (2010).
139 groundWork (2010c: 3).
140 groundWork et al. (2010: 1).
141 Ibid.
142 groundWork (2010c: 3).
143 groundWork (2010f). World Bank money for Eskom will amplify South Africa's energy, climate and poverty crises. 16 February. http://www.groundwork.org.za/Publications/EskomFinalDocs/WBEskomcritique%20pdf.pdf
144 groundWork (2010g: 5).
145 World Wildlife Fund 2010a. WWF responds to approval of World Bank loan. 8 April. Requested directly from WWF South Africa, for a copy contact snobula@wwf.org.za.
146 groundWork et al. (2010: 2).
147 Greenpeace Africa. 2010a, March 31. Statement from Greenpeace Africa on the World Bank's Proposed $3.75 Billion Loan to Eskom. www.greenpeace.org/africa/en/press-centre-hub/press-releases/statement-from-greenpeace-africa-on-the-world-banks-proposed-375-billion-loan-to-eskom/
148 groundWork (2010g: 4).
149 World Wildlife Fund. 2010b. More froth than fury. Business Day. Letter to the editor from Richard Worthington, manager of the Climate Change Programme, Living Planet Unit, WWF South Africa. 10 March. http://www.businessday.co.za/articles/Content.aspx?id=95836
150 Africa Action (2010b).
151 groundWork (2010h).
152 Wri (2010).
153 groundWork 2010d. Eskom and World Bank Sowing Seeds of Destruction." GroundWork Quarterly Newsletter 12 (1). 1 March. http://www.groundwork.org.za/Newsletters/March2010WEB.pdf
154 groundWork (2010f).
155 Wri (2010).
156 groundWork (2010g: 2).
157 WWF (2010a).
158 groundWork (2010e: 2).
159 groundWork (2010f).
160 Wri (2010).
161 WWF (2010a).
162 Africa Action.
163 groundWork (2010b).
164 Rafey and Sovacool (2011: 1141–1151).
165 Pegels, A. 2009. Prospects for Renewable Energy in South Africa: Mobilizing the Private Sector. Discussion Paper. Bonn: German Development Institute/Deutsches Institut für Entwicklungspolitik; 23/2009.
166 Ferrey (2009: 17–28).
167 Winkler, H., A. Hughes, and M. Haw. 2009, November. Technology Learning for Renewable Energy: Implications for South Africa's Long-Term Mitigation Scenarios. Energy Policy 37 (11): 4987–4996.
168 Baker, and Sovacool (2017: 1–112).

169 Jenkins, J. 2010, March 23. Without Affordable Clean Alternatives, South Africa Turns to Coal. http://thebreakthrough.org/blog/2010/03/without_affordable_clean_ alter.shtml
170 Meyer, E. L., and K. Odeku. 2009, Winter . Climate Change, Energy, and Sustainable Development in South Africa: Developing the African Continent at the Crossroads. Sustainable Development Law & Policy 9 (40): 53, 74–75.

6

SOCIOTECHNICAL IMAGINARIES

Smart meters and the public in the United Kingdom

Smart meters may sound bland but they are a legitimately useful tool for better measuring (and perhaps reducing) energy consumption in homes, and even offices, businesses, and industrial facilities. In the United Kingdom, smart meters are replacing traditional meters for monitoring gas and electricity consumption. Smart meters measure, store and share data on digital networks. It is hoped that smart meters will transform energy use leading to demand reductions and a shift in peak demand. Smart meters are expected for instance to make a significant contribution in meeting efforts to decarbonize households. Moreover, replacing old unreliable traditional meters with new smart meters can result in more accurate readings and billing, shifts in peak energy demand and/or to when it is lower carbon, and improved reliability and security, including reduced power outages.

The Smart Meter Implementation Program (SMIP) lays the legal foundation to offer a smart meter for electricity and for natural gas to *every* home and small business in England, Scotland, and Wales by the end of 2020. It represents the United Kingdom (UK) government's "flagship energy policy" and involves installing a combined 104 million pieces of new equipment at a total cost of at least £11 billion.[1] Although the expected costs and benefits of the rollout are debated, Lewis and Kerr have argued that the SMIP is "by far the most complex" and also "costliest" smart meter program in the world, as well as the largest government-run information technology project in history.[2] Smart Energy GB, the "voice" of the smart meter roll out, framed it as "the biggest behavioral change program that this country has seen" and "the biggest national infrastructure project in our lifetimes."[3] This makes the UK an obvious frontrunner in the deployment of smart meters, and as we will see below, they also favor a unique approach that couples smart meters (for both gas and electricity) with in-home displays (IHDs) giving real-time information to users.

In this chapter, drawn from previous research with Sabine Hielscher,[4] I explore the SMIP from an unusual perspective: that of sociotechnical imaginaries. The chapter asks: how are smart meters (and their rollout) being discussed and framed with the popular news media? And: How do these imaginaries change over time?

To provide an answer, Sabine conducted a systematic review and content analysis of popular newspaper and tabloid articles published in the UK media from 2006 to 2016. The result of this search revealed hundreds of documents Sabine and I then analyzed to assess the prevalence of utopian and dystopian narratives, which I divide into nine distinct imaginaries. Four of these—"empowered consumers," "energy conscious world," "low-carbon grid," and "inclusive innovation"—depict smart meters as a harbinger of positive social change. Five of these—"hacked and vulnerable grid," "big brother," "costly disaster," "astronomical bills," and "families in turmoil"—frame smart meters as destructive, negative forces on society. I conclude with insights about what such competing imaginaries mean for energy and climate policy as well as expectations about sustainability transitions.

In proceeding on this path, the chapter aims to make multiple contributions. Various research has explored the fantasies or imaginaries associated with new sources of energy supply—notably clean coal,[5] hydrogen fuel cells[6,7] nuclear power,[8] advanced biofuel[9,10] bioenergy[11] and solar energy,[12] to name a few. However, none has of yet explored energy demand or end-use devices such as smart meters. Moreover, research has tended to emphasize visions or imaginaries shared by technical experts or elites. This includes the scientists working as part of the Intergovernmental Panel on Climate Change[13] or advising the creation of the Montreal protocol for chlorofluorocarbons,[14] or physicists at the International Atomic Energy Agency (see Chapter 2) and academic researchers in the hydrogen community (see Chapter 3). Less research has emphasized visions shared or promoted by the popular news media or members of the public themselves.

The chapter first defines and conceptualizes a "smart meter" and then discusses the dynamics of the UK smart meter rollout and introduces the conceptual framework of sociotechnical imaginaries. It then explicates its research design and empirical strategy before moving to the primary discussion of the nine smart meter imaginaries.

Conceptualizing and defining smart meters and the UK smart meter rollout

Although there is no universally accepted definition of what constitutes a "smart" energy or gas meter, the phrases "advanced meter" or "smart meter" can refer to a bundle of different systems including net meters, digital meters, automated meters, interval meters, new meters, retrofitted meters, two-way communication devices, monitors and displays, and more.[15] Purpose and functionality generally distinguish "smart" meters from "dumb ones," that is, "smart" meters communicate electronically and via a network.[16]

Interestingly in UK policy documents, a range of terminology (such as "new types of meters" and "smart meters") was used until smart meters became a more commonly used term from 2006.[17] In the UK, smart meters have come to mean meters that can both measure and store data at specified intervals, and act as a node for communications between supplier and consumer via automated meter management (AMM). An "in-home display," or IHD, refers to the device or monitor that connects with the smart meter and provides consumers with information about their energy consumption and costs.[18] The term "advanced metering infrastructure," or AMI, is meant to encompass the entire system of associated communications and infrastructure involved in supporting and facilitating smart meters.[19] As Figure 6.1 indicates, when one focuses on the entire web of AMI rather than only the meter itself, the SMIP involves the simultaneous conversion of smart electricity meters, gas meters, IHDs, and wireless area networks, as well as households, data and communications companies, service users, and electricity and gas utilities.

Smart meters, or devices similar to them, have a fairly long history in the UK. A commitment to modern smart meters in domestic buildings, and the SMIP as we know it today, started to slowly emerge from the 2000s with a renewed focus on energy security and climate change, and the opening up of supply competition for households, plus the development of information and communication technology (ICT), as Table 6.1 summarizes. Driven by the European Union Energy End-Use Efficiency and Energy Services Directive 2006/32/EC in 2008, Gordon Brown's government announced its decision to rollout smart meters, including display units to all households by the end of 2020. The announcement was made before the results of the government backed pilots to assess the benefits of smart meters, such as the Energy Demand Research Project from 2007 to 2010, were collated and the impact assessments were fully completed.[21] The government also created a legal framework for the rollout by implementing regulatory changes using powers conferred on the Secretary of State by the Energy Act 2008 and later the Energy Act 2011.

Since 2001, a substantial regulatory apparatus has been slowly implemented, setting in motion the SMIP, starting off with various consultations, pilots, and trials (2001 to 2010) before moving to a policy design stage (July 2010 to March 2011), and followed by the foundation stage (March 2011 to 2016), and the rollout (November 2016 to present). In terms of implementing the rollout, the UK government has made energy suppliers responsible. The control of the smart metering communication system has been delegated to a licensed Data and Communications Company (DCC) who forms contractual ties with the suppliers. Energy suppliers are intended not only to install the technologies (i.e. smart meter, in-house display and digital communication hub) but to convince householders to adopt them as outlined in a Consumer Engagement Strategy. Adoption is expected to occur in "every home," although national planners have more recently emphasized that the SMIP is only for households "that want one," or that "everyone will be offered the opportunity to upgrade to a smart meter," giving them the option of not participating.

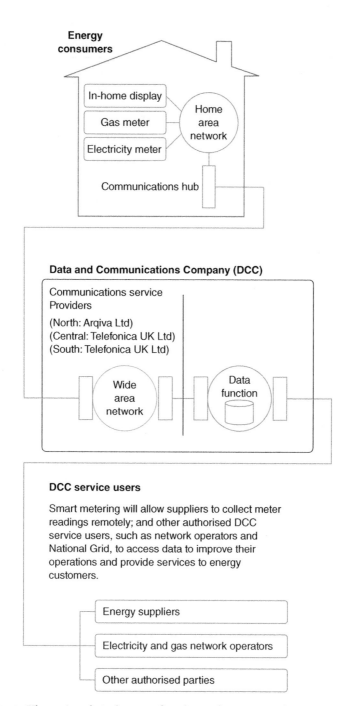

Energy consumers

In-home display

Gas meter

Electricity meter

Home area network

Communications hub

Data and Communications Company (DCC)

Communications service Providers

(North: Arqiva Ltd)
(Central: Telefonica UK Ltd)
(South: Telefonica UK Ltd)

Wide area network

Data function

DCC service users

Smart metering will allow suppliers to collect meter readings remotely; and other authorised DCC service users, such as network operators and National Grid, to access data to improve their operations and provide services to energy customers.

Energy suppliers

Electricity and gas network operators

Other authorised parties

FIGURE 6.1 The sociotechnical system for advanced metering infrastructure and smart meters in the United Kingdom

Source: Author, derived from Sovacool, B. K. et al. 2017, October[20]

TABLE 6.1 Chronological timeline of the smart meter implementation program in the UK

Date	Events	Phases
2001	Smart Meter Review Group	
2006	Ofgem consultation on smart meters	
January 2006	EU Energy End-Use Efficiency and Energy Services Directive	
March 2006	Government announced energy suppliers a pilot study of feedback devices such as "smart" energy meters	
March 2006	Ofgem review: International experience of smart metering	
August 2007	Government consultation: Views on full roll-out of smart meters	*2001 to 2010 Consultations and Pilots*
2009	DECC Impact Assessment of GB-wide rollout	
May 2009	Government consultation different implementation models	
July 2009	European Commission Directive 2009/72/EC	
December 2009	Decision to have smart meters in all home in Great Britain by 2020	
2010	Prospectus for the Smart Metering Implementation Programme	*2010 to 2011 Policy Design*
2011	Announcement: Start mass rollout in 2014—Completion in 2019	
2011	Suppliers start to install "smart meters"	
June 2011	Energy Demand Research Project final analysis—Ofgem	
July 2011	First National Audit Office carries out review	
April 2012	Consumer Engagement Strategy Consultation	*2011 to 2016 Foundation*
May 2012	Monitoring and evaluation strategy	
December 2012	SMETS 1 meters begin to be offered	
July 2013	Launch of Smart Energy GB	
September 2013	Appointment data and communications company by DECC	
November 2014	Delay of smart meter rollout	
March 2015	Smart metering Early Learning Project: Synthesis report	
August 2015	DECC Consultation on amendment of Smart Meter In-home Display license conditions	
November 2016	Start of main rollout	*2016 to 2020 Main Installation*

Source: Author, modified from Hielscher, S., and P. Kivimaa 2018[22]

Figure 6.2 shows that 10.8 million smart meters have been installed as of early 2018,[23] corresponding to about 19 percent of the target number of 56 million. Now, to make sure that the SMIP meets targets for the remainder of the program, suppliers must complete installations at a rate of about 40,000 per *day*. The rate of installation has increased dramatically over the last few quarters as suppliers have ramped up their installation capacity and this trend looks likely to continue.

Sociotechnical imaginaries

To analyze the broader visions and narratives around the SMIP, I relied on a conceptual approach known as sociotechnical imaginaries. The roots of this approach can be traced back to political and social theory (notably Anderson,[24] Appadurai,[25] and Taylor[26]) as well as science and technology studies.[27,28] At a very general level, these works all investigate how given social systems cohere and absorb internal tensions and contradictions, how the practices of collective imagination can resolve conflict and mold consensus, and how imaginaries can come to constrain and shape behavior.

However, despite both the utility and novelty of such themes, Jasanoff warns that:

> Science and technology have been involved in efforts to reimagine and reinvent human societies for close to two hundred years. Social theory,

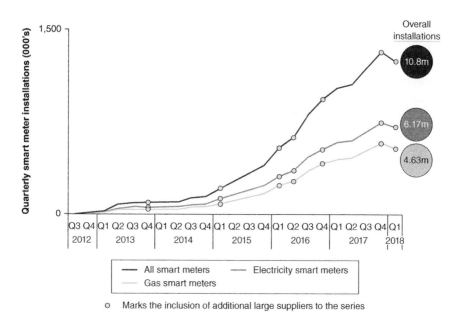

FIGURE 6.2   Smart meter installations in the UK for large energy suppliers, 2012 to 2018
Source: BEIS 2018

however, has yet to embrace this key dimension of modernity and to acknowledge the centrality of these two institutions in constructing the futures toward which we direct our presents.[29]

Building on such earlier work, Jasanoff and Kim define a sociotechnical imaginary as "collectively imagined forms of social life and social order reflected in the design and fulfillment of nation-specific scientific and/or technological projects."[30] A more recent definition refers to "collectively held, institutionally stabilized, and publicly performed visions of desirable futures, animated by shared understandings of forms of social life and social order attainable through, and supportive of, advances in science and technology."[31] Imaginaries thus reflect the capacity of people (and institutions) to imagine futures, and such imaginaries can offer a resource that mobilizes resources and helps define agents (and technology).

Sociotechnical imaginaries research often prioritizes a focus on the three "Ms" of materiality, meaning, and morality. In terms of materiality, it emphasizes that technologies themselves can act as physical forces that constrain action or enhance an experience. In terms of meaning, technologies can provoke radically different reactions from different stakeholder groups. And, lastly, in terms of morality, technologies can often have negative, pejorative impacts on society. Jasanoff notes that therefore, the theory "cuts through the binary of structure and agency: it combines some of the subjective and psychological dimensions of agency with the structured hardness of technological systems, policy styles, organizational behaviors, and political culture."[32] Jasanoff later adds that imaginaries "occupy a hybrid zone between the mental and the material, between individual free will and group habitus, between the fertility of ideas and the fixity of things."[33] Pickersgill writes that "sociotechnical imaginaries are shown to be salient in structuring anticipatory discourse, and represent a key target for social scientific intervention in such debates."[34] Jasanoff and Kim also suggest that "sociotechnical imaginaries are powerful cultural resources that help shape social responses to innovation."[35]

Imaginaries differ from pure discourse analysis because the former usually focuses on language (see Chapter Three) whereas the latter emphasizes action and performance with materialization through technology. Imaginaries are not the same as policy agendas or frames either (see Chapter Four), as imaginaries are less explicit and accountable. Nor are they the same as narratives, which are usually extrapolated from past events and serve explanatory or justificatory purposes.[36] Imaginaries instead are instrumental and futuristic, they project visions of what is good, and worth attaining (and also, as this chapter adds, dystopias worth avoiding).

In their examination of the imaginaries surrounding nuclear power in South Korea and the United States, Jasanoff and Kim note that they can differ meaningfully in how they define risk, policy foci, controversies, stakeholders, forms of closure, and even underlying epistemology.[37] The American sociotechnical imaginary for instance was identified as perceiving the technology's benefits as limitless whereas risks are framed as discrete and manageable; the Korean imaginary did not. Other imaginaries may focus not on the introduction of some new technology but instead phasing out an undesirable one. Felt for instance has noted a series of imaginaries of "Austria being free" of specific kinds of technology such as nuclear energy, biotechnology, or nanotechnology.[38] Although they do not use an imaginary framing, I also see compelling assessments arguing for the disuse, phasing out, or un-invention of nuclear weapons[39] or fossil-fuel based energy systems.[40]

Felt writes that there is a temporal element to imaginaries as well, with some (in Austria at least) following a general progression or sequence of four stages.[41] The first stage involves a *process of assembling* a case for (or against) a particular technology or vision, opening it up to the public. The second stage involves *multiple rehearsals* to both anchor the issue in different public arenas and to reach a stable outcome. Successful rehearsal, especially when it involves iconic pictures, slogans, and stories, leads gradually to *stabilization*, which creates a standardized history resistant to competing interpretations. The final stage is *transference*, when the stabilized imaginary becomes part of a collective identity and then integrated into other technological debates.

Viewing new technologies and the process of diffusion as a sociotechnical imaginary has advantages. It focuses on imagination, and the common narratives that vibrant societies often have about who they are, where they have come from, and where they are headed, usually through the interplay of positive and negative imaginations—utopia and dystopia. It rejects political determinism and the idea that politics or action is always purposively rational. The approach emphasizes a performative element to technology, that "unlike mere ideas and fashions, sociotechnical imaginaries are collective, durable, [and] capable of being performed; yet they are also temporally situated and culturally particular."[42] By exploring how sociotechnical projects travel from imagination and conception to realization, the theory helps uncover the process of extension, where particular narratives or ideas gain traction, acquire strength, and cross scales.[43] Imaginaries remind us also that the public (or diverse publics) can not only be suppliers of visions and narratives but mass consumers of them. Lastly, imaginaries have an inherently subjective element, making them unique: one person's utopia can be another's dystopia. Utopias have a central function in society as ensuring that humans strive to create the best possible society (see Chapter Two) but they can also offer sobering criticism of existing social patterns.[44] Imaginaries can thus serve the interests of both conservatives and radicals, crossing political, ideological, and socio-technical divides.

Research design and empirical strategy

To identify public imaginaries, the main data source for the chapter's analysis consisted of articles from UK national newspapers, containing the keyword "smart meter" at the start of the article and that were listed in the Lexis Nexis database. The purpose was to cover imaginaries about the rollout of smart meters since the introduction of the European Union Energy and End-Use Efficiency and Energy Service Directive and throughout the policy design and foundation stage until the first few months of the start of the mass-rollout. Therefore, the analyzed period was from January 2006 to May 2017. The resulting corpus included 283 publications containing 500 words or more. After closely reading the documents for their relevance to the smart meter rollout, 250 articles (summarized by Table 6.2) were chosen for analysis. There has been a steady increase in the amount of articles being published until 2014, including one peak in 2009, when the UK government announced the rollout (see Figure 6.3). Since 2014, there has been a steady increase in newspaper coverage.

Admittedly, the term "smart meter" might have limited the search, not including less direct references to these technologies, and broader discourses, for instance, around "smart energy," "smart cities," "smart homes," and "smart grids." This was done to focus the examination on imaginaries surrounding the smart meter rollout specifically, to be able to explore which broader themes are attached to the technologies and rollout in the UK, and to see the role they have taken in public discussion. Both broadsheet and tabloid press has been included in the analysis. Although broadsheet press is argued to be the "quality press,"[46] the analysis of tabloid press is of high significance when researchers are interested in mass public discourse, considering that large proportions of the population read tabloid newspapers.[47]

The utopian dynamics of smart meter imaginaries

Qualitative coding of the final sample of 250 newspaper articles revealed four prevalent, recurring utopian imaginaries, as Table 6.3 summarizes. The most frequent imaginary presented (n=116) was one of "empowered consumers" followed by "energy conscious world" (n=104), "the low-carbon grid" (n=44), and "inclusive innovation" (n=12). As Table 6.3 also summarizes, such imaginaries have different discursive elements, symbolic cues (recurring phrases), and intended institutions and audiences.

Empowered consumers

The imaginary of "empowered consumers"—the most prevalent positive one from the sample of media articles—frames smart meters as a pathway towards more accurate bill reading, easier switching for suppliers, and—best of all—making money in the home. The coalition of actors and institutions involved in this imaginary centre on government bodies such as DECC and Ofgem, suppliers such as British Gas, and consultancies and civil society institutions.

As material from the sample indicated, *The Guardian* wrote that smart meters eliminate the need for customers to "stay at home for meter readings or to receive over-estimated bills" and can enable smart meters to "allow consumers

TABLE 6.2 Overview of smart meter newspaper content by source, 2006–2016

	2006	2007	2008	2009	2010	2011	2012	2013	2014	2015	2016	2017
Broadsheet												
The Guardian.com		1		4	1	5	2	3	1		1	
The Guardian			1	6	1			1		4	2	5
The Observer			1	2	1		2	1				
The Sunday Times			2	2	2		1		2	3	4	
The Times			1	10	5	6	2			2	3	3
Independent.co.uk						4	2		2			
The Independent	1		1	3	1						2	
The Independent on Sunday			1			2	1					
The Daily Telegraph		2		2	2		3	2	1	1	7	5
The Sunday Telegraph					1		1		1	4		
Total	**1**	**3**	**7**	**29**	**14**	**17**	**14**	**7**	**7**	**14**	**19**	**13**
Tabloid												
Daily Mail		2	1	1	1	1	2	1		1	1	1
The Mail on Sunday							1		1	2		
Mail Online							2	2	8	5	8	9
Telegraph.co.uk					1		3	8	2	9	6	4
Mirror.co.uk											3	
The Mirror				1								
Daily Mirror										1	4	
The Sun										1	1	
Express Online								1	1	2	1	2
The Express				1		1						
Sunday Express				1						1		
Total	**0**	**2**	**1**	**4**	**2**	**2**	**8**	**12**	**12**	**22**	**24**	**16**

to feed the electricity they generate through solar panels into the grid"). *The Independent* concurred and noted that with smart meters "energy bills are likely to fall by more than £20." *The Mail Online* suggested that "the national smart meter roll-out is the essential transformation of the technology we use to buy energy … It will create newly empowered consumers, and increase trust in those who sell us

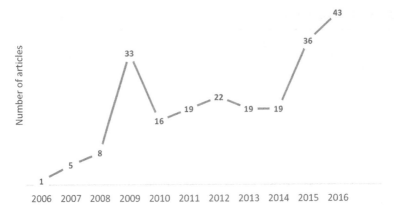

FIGURE 6.3 Number of UK newspaper articles discussing smart meters, 2006 to 2016 (Total n=250)

Source: Author, based on Hielscher, S., and B. K. Sovacool 2018, September.[45] Note: 2017 has been excluded from the figure because the amount does not represent the full year

gas and electricity." *The Telegraph* also agreed when it wrote that "when it comes to energy management, it's truly empowering."

Between 2006 and 2009 in particular, several storylines were part of the imaginary of "empowered consumers" that were linked to a diverse set of actors (e.g. think tanks and government ministers). The possibility of introducing more accurate energy household bills was the most stable and permanent storyline alongside two more contested storylines around choosing tariffs that suit people's needs and saving money on energy bills. From 2012 onwards, the storyline changed to reflect more sanguine assessments, noting that smart meters would "hopefully give value for money" or that they "should actually save money" (*The Daily Telegraph*).

Energy conscious world

The "energy conscious world," the second most frequent utopian imaginary, focused less on empowerment and moneymaking, and more on better comprehension of energy patterns and consequent improvements in energy efficiency. This imaginary involved mostly energy suppliers and government ministries with some civil society groups such as the Energy Savings Trust.

For instance, *The Independent* wrote that smart meters "will help you grasp the power-consuming patterns of your domestic lifestyle" and that "those who use a smart meter would be more inclined to make energy saving decisions such as insulating their lofts, turning down their thermostats and switching appliances off rather than leaving them on standby." *The Guardian* emphasized that "Sophisticated monitoring systems as part of a smart grid will allow the network to match demand with supply in a far more efficient way." The *Times* wrote that "By encouraging households to run appliances such as dishwashers and washing

TABLE 6.3 Summary of four utopian imaginaries surrounding smart meters in the UK

Imaginary	Frequency	Description	Symbolic cues	Coalition of actors and institutions
Empowered consumers	n=116	Smart meters facilitate more accurate bill reading, easier switching between suppliers, prosumption, and making money	"money-saving" changes; "cut power bills," "end of debt," "accurate bills," "help the fuel-poor," "make deals," "no longer science fiction," "forget estimated reading"	Energy Saving Trust, Green Alliance, Sustainability First, Northern Ireland Electricity: Cut power bills Centrica: cuts energy use, micro-generation British Gas: Open the door to energy efficiency measures Hendry: save money Ofgem: Saving of £134 a year DECC: Help hard-working consumers
Energy conscious world	n=104	Smart meters enable consumers to grasp their energy patterns, saving energy and making homes more energy efficient	"Wasting electricity at home;" "energy-conscious world," "greater environmental awareness;" "save the planet;" "boost energy efficiency;" put inefficiency "into the history books;" "do your bit" to save the environment	Ofgem, DECC and Energywatch: gain clearer information on energy use Energy Savings Trust: A good nudge Energy Retail Association: Consider gas and electricity Moneysupermarket. com: curb unnecessary consumption National Grid: Balance supply and demand British Gas: save energy and money
The low-carbon grid	n=44	Smart meters promote the decarbonisation of electricity and gas	"cut carbon emissions," "low-carbon energy future," "green revolution," "combat climate	National Grid: Smart grid Government ministers: Help to meet carbon targets

(Continued)

TABLE 6.3 (Cont.)

Imaginary	Frequency	Description	Symbolic cues	Coalition of actors and institutions
			change," "pave the way for renewables," "intelligent networks"	DECC and Ofgem: reduce carbon intensity
Inclusive innovation	n=12	Smart meters enhance industrial strategy and economic competiveness	"boost competition," "innovative energy markets," "exciting new innovation," "smart home for everyone"	BEIS: industrial strategy Smart Energy GB: smart innovation for every home

Source: Author, based on Hielscher, S., and B. K. Sovacool 2018, September[48]

machines at off-peak times, the meters will also help to spread electricity demand across the day."

From 2012 onwards, the actors and institutions endorsing this imaginary no longer included consumer protection organizations and non-governmental organizations but mainly energy companies (in particular, British Gas) and the UK government, and from 2015 onwards also the national campaign for smart meter rollout. At the time, *The Sunday Telegraph* quoted the national campaign for the smart meter rollout, arguing, "people who already have a smart meter receive 'easy access to clear information … about how much gas and electricity they are using'." Disagreements in this storyline appeared in 2013 and 2014, when energy companies argued for "removing the obligation of the mandatory requirement if in-home displays significantly reduce the costs of the smart meter program to consumers" (*The Telegraph*, 2013) and the energy department responded "In-Home-Displays will give consumers easy access to information on their energy consumption in pounds and pence, to help them manage and control their energy use" (*The Telegraph*, 2013).

A less prevalent storyline within the "energy conscious world" imaginary derived from government actors, energy companies, and later on the national marketing campaign associated with the rollout. It was grounded in envisioning the possibilities of smart meters enabling the efficient management of energy resources through better management of supply and demand, and reducing peak demand. For instance, in 2009, *The Guardian* wrote "smart meters help contribute towards more efficient—and greener—management of the electricity grid," and in 2011 emphasized that "sophisticated monitoring systems [i.e. smart meters] as part of a smart grid will allow the network to match demand with supply in a far more efficient way."

The low-carbon grid

The imaginary of the "low-carbon grid" explicated the carbon and environmental credentials of smart meters. The coalition here involved almost exclusively government actors and energy suppliers.

For instance, *The Independent* wrote that the smart meter program remains "a central component of the UK's low-carbon energy future." *The Guardian* published that the program is the "vision for a low carbon Britain" and that "we need a smart grid in order to cope with a substantial increase in the proportion of our electricity that comes from intermittent clean energy sources such as solar and wind" as well as better integration with electric vehicles. *The Telegraph* wrote that smart meters "make us use less energy and contribute towards a low-carbon future, along with wind turbines and other renewables" *The Times* wrote that "smart meters were an important part of British plans to cut carbon emissions" and *The Daily Mail* that "Ministers believe this will encourage families to use less, helping the UK meet EU carbon reduction targets."

This imaginary was further built upon when linking smart meters to enabling a smart grid; "Smart grids will help manage the massive shift to low carbon electricity such as wind, nuclear and clean fossil fuel" (*The Guardian*). After regular consultations around the costs of the rollout and increasing energy bills, the Science and Technology Committee (SCT) was referred to in the *Express Online*, arguing that "it would be 'easy to dismiss the smart meter project as an inefficient way of saving a small amount of money on energy bills' but evidence suggested there were major national benefits such as a smarter and more secure grid."

Inclusive innovation

This imaginary of "inclusive innovation" emphasized the contributions the smart meter rollout would offer for industry, innovation, and economic competiveness. The coalition here involved primarily government ministries and affiliated actors.

For example, *The Times* wrote that "next-generation smart meters are paving the way for even more exciting innovations to conserve energy." The *Daily Mirror* wrote:

> even if you're not one of those people who have fully embraced smart devices, there's a good chance that within a few years you could well be. Because, slowly but surely, our lives are being transformed by the 'Internet of Things'; the idea of linking devices and appliances to the world wide web.

The Mail Online wrote that:

> the smart city is an alluring vision of the future, in which civic technology such as traffic lights, smart meters for utilities and public transport could all

be connected and feed back invaluable data online, step on the road to smart city.

From 2011, the imaginary was mainly grounded in developing novel services "to boost competition and innovation in the energy market" (*The Times*). *The Independent* (2014) wrote, "they [smart meters] also create opportunities for innovative new services to be developed." From 2014, these storylines were linked to smart home and smart city. For instance, *The Guardian* (2015) noted that "the smart city is an alluring vision of the future, in which civic technology such as traffic lights, smart meters for utilities and public transport could all be connected and feedback invaluable data online."

The dystopian dynamics of smart meter imaginaries

Visions, narratives, and imaginaries about smart meters, interestingly, were not only positive. Our sample also revealed strong negative dystopian imaginaries. In order of the most prevalent, these included "costly disaster" (n=103), "astronomical bills" (n=79), "big brother" (n=51), "the hacked and vulnerable grid" (n=16), and "families in turmoil" (n=9). Table 6.4 offers a summary of these imaginaries.

Costly disaster

The most prevalent dystopian imaginary focused on the smart meter rollout as a "costly disaster," one that will only retard innovation and one where technology will become rapidly obsolete. The coalition here involved government watchdogs such as the National Audit Office, energy consulting firms, and some suppliers critical of the rollout such as EDF, Scottish Power, and Npower.

As *The Times* wrote, "a final decision [on smart meters] had been delayed until November because the Government is not yet convinced that the meters will be cost-effective" and "the Government has underestimated the cost of a nationwide rollout of smart meters by as much as £6.4 billion." The *Daily Telegraph* wrote that the smart meter program could result in an "embarrassing and costly failure" and that it was "at risk of becoming a fiasco." *The Observer* warned of "stupendous cost" and *The Guardian* cautioned that the program may be "a costly failure." *The Guardian* later wrote that "smart metering is firmly on course to be the next big UK Government IT disaster" and that "rather than see reductions in costs, I fear that most consumers will pay more, the most vulnerable people will be penalised and no energy will be saved at all." *The Independent* reported that "Consumers will have to pay for smart meters even though they might already be out of date." *The Sunday Times* critiqued the rollout for being an "IT disaster waiting to happen" and "over-engineered and mind-blowingly expensive."

A particularly powerful coalition of actors, who had shared storylines connected to this dystopian imaginary, emerged from 2011 onwards in the

TABLE 6.4 Summary of five dystopian imaginaries surrounding smart meters in the UK

Imaginary	Frequency	Description	Symbolic cues	Coalition of actors and institutions
Costly disaster	n=103	The smart meter rollout will be a publicly funded technology disaster with inconclusive benefits and outdated technology that will retard innovation	"cost more than expected," "IT disaster," "embarrassing and costly failure," "energy reduction inconclusive," "white elephant," "financial fiasco," "unexpected delays," "overspend," "most costly in the world," "true costs," "technological obsolesce," "immature technology," "nightmare to install," "faulty meters," houses trapped in "billing limbo"	National Audit Office: Costs could escalate Mott Macdonald, Industry, government, Centrica (British Gas), ERA, BERR: unconvinced cost effective rollout EC (EDF, Scottish Power, Npower): call for review of rollout, Ernst & Young: underestimation of costs Datamonitor: immature technology
Astronomical bills	n=79	Consumers are forced to install and pay for smart meters which will only lead to higher tariffs and bills	"households at risk of being ripped off," "ramp up costs," "bamboozled by tariffs," "sales tricks," "extraordinary bills," "costs will be passed on," "limited evidence of cost savings," "people will pay more," "consumers are paying"	Consumer groups: Consumers shoulder heavy costs McKinnon and Clarke: EC clutching onto straws to create positive story Huhne: can increase price of electricity Consumer Focus: concerned about poorest consumers— only middle-class benefits Inenco (energy consultancy): force prices up PAC: Set to cost consumers more than they save Energyhelpine: keeps consumers hostage

(Continued)

TABLE 6.4 (Cont.)

Imaginary	Frequency	Description	Symbolic cues	Coalition of actors and institutions
Big brother	n=51	Smart meters erode privacy protections	"the spying utility," "snooping on peoples' lives," "invasion of privacy," "captive consumers," "permanent window on private life," "spy on the home," "malpractice," "mistrust"	Datamonitor: privacy issues Privacy campaigners: Data might be used to target consumers Consumer groups: Power companies micro-manage electricity supply Big Brother Watch: Massive intrusion into what goes on in people's lives, increase surveillance of society Privacy International: risks of identity theft, real time surveillance, unwanted publicity, profiling or targeting for commercial purposes
Hacked and vulnerable grid	n=16	Smart meters can facilitate criminal theft of data or sabotage of the electricity or gas network	"system security," "hacking and manipulation," "scammers," "information insecurity,"	DECC: System security crucial issue Privacy campaigners: public does not realize how much detail can be gleaned CESG: The security spec is a sticking plaster
Families in turmoil	n=9	Smart meters disrupt family routines and can lead to the policing of activity by children	"daddy is destroying the planet," "nagging about energy waste," "people bickering over energy use," "rise of household tensions," "the eco police"	Parents: nagging family members, complaining children Consumer groups: added stress on family routines

Source: Author, based on Hielscher and Sovacool 2018: 978–990.[49] Note: ERA = Energy Retail Associations. BERR = Department for Business Enterprise and Regulatory Reform. EC = Electricity Commission. EDF = Électricité de France. PAC = Public Accounts Committee. CESG = National Technical Authority for Information Assurance

newspapers. The coalition consisted of several parliamentary committees, consumer protection organizations, and academics pointing to uncertain costs and benefits, and warning that costs might escalate. The coalition, in particular, drew on two publications: one produced by the National Audit Office,[50] cautioning that the rollout "could end up costing more than it is supposed to save" (*The Times*) and another written by a parliamentary committee.[51] A consumer protection organization referred to the latter report, arguing in one of the articles "that the roll-out might end in embarrassing and costly failure" (*The Daily Telegraph*).

Astronomical bills

The "astronomical bills" imaginary emphasized the financial setbacks of the rollout, especially in situations where smart meters can enable companies to levy higher tariffs or disconnect non-paying customers. The primary coalition here involved consultancies and consumer groups.

For example, *The Daily Mail* writes that "homeowners could face a bill of up to £400 to install hi-tech electricity and gas meters under Government plans" and that "nothing can be done to stop them making customers pay." The *Guardian* wrote that "homeowners ... have to shoulder heavy costs for the new meters." The *Daily Telegraph* was concerned about "the impact the measures would have on the poorest consumers." The *Telegraph* wrote that "bills will rise over this decade to pay for the installation programme, while critics say the savings are uncertain." *The Sunday Times* stated that "government officials admitted that this cost would largely be passed on to consumers through a levy on bills" and there is no "cap on costs, which might have been one of the ways you could have protected the consumer." *The Observer* wrote that "the utilities love the technology because it will enable them to disconnect consumers remotely who don't pay their bills." The *Daily Mail* noted that "consumer groups have also warned that the devices will allow suppliers to cut off energy at the 'flick of a switch' without even having to enter people's homes." The *Mail Online* wrote that:

> It means electricity and gas used in the evenings could cost 99 per cent more than at other times—penalizing everyone cooking family meals, watching popular TV shows and heating their homes on chilly winter evenings. Higher charges will also apply in the morning when people are most likely to be taking baths and showers and having the central heating on.

Thus, the "astronomical bills" imaginary emphasized the financial setbacks of households connected to the rollout, especially in situations where smart meters enabled energy companies to charge more at peak times through new tariffs, keep savings and fail to pass savings on to consumers, or disconnect non-paying customers. The dystopian imaginary focused on consumers having to subsume the costs of smart meters without any tangible mechanism being in place to gain some of the benefits, which would only lead to higher energy bills for households. Contestations associated with this dystopian imaginary consisted of debating the probability of consumers reducing their energy bills through using less energy in the home, the readiness of consumers choosing tariffs that help them reduce their bills, the ability to switch more easily between suppliers in order to get the cheapest deal, and the likelihood of energy companies' savings being passed on to consumers through lowering their bills. The storyline slightly weakened from 2012 onwards: actors added that a competitive market would make sure consumers did not pay too much, "the department and energy companies claim significant savings will be made in the long-term. The meters should reduce energy suppliers' costs, which, the government hopes, will mean more competitive tariffs for consumers" (*The Sunday Times*).

Big brother

This dystopian "big brother" imagery emphasized how smart meters would enable "spying utilities" and "snooping" to the point where privacy was invaded. This coalition involved mostly consumer groups and civil society organizations.

As an example of this narrative, *The Sunday Times* wrote that smart meters "will reveal when people are at home, what sort of appliances they are using and even indicate their diet … Privacy campaigners say the public does not realize how much detail can be gleaned from the meters" and that "this could be used by the government, for example, to tell if somebody who is claiming benefits has bought a television, or to tell how many people are living in a particular home." *The Sunday Telegraph* warned that "energy companies could hold customers to ransom by making it difficult to switch." *The Guardian* published a story noting that "unless we are very careful, we will see Big Brother taking over our homes as power companies get to micro-manage our energy supply and are given complete access to information about how we live." *The Independent on Sunday* cautioned that "smart meters could open Pandora's box: The big six energy companies have, according to Consumer Focus, an 'appalling track record' over mis-selling to customers." The *Daily Telegraph* wrote that:

> there is a risk that the customer will feel even more captive and question whether there are any real benefits unless the suppliers are willing to add a wider range of services than simply using the 'spy' in the home to register electricity and gas consumption.

Hacked and vulnerable grid

The "hacked and vulnerable grid" imaginary noted how smart meters could result in an electricity system subject to hacking or vulnerable to criminals. This coalition primarily involved a mix of privacy campaigners and government groups.

As an example of this narrative imagery, *The Guardian* warned that:

> the threat of internet viruses infecting home computers and mobile phones is something we have all learned to live with, but soon many homes' energy supplies could face similar risks. Security experts say smart meters are also potentially vulnerable to hacking and manipulation. Moreover, the masses of data they produce on energy consumption habits could prove valuable to thieves, scammers, or unwanted telemarketers.

The Mail Online wrote that "criminals could break into the system and try to switch off the supply to millions of homes at once, leaving the national grid crippled." They also wrote that:

> intelligence chiefs have warned that plans to install smart energy meters in every house will leave families vulnerable to terrorist attacks. According to the Government's listening agency GCHQ, the plans will create a 'strategic vulnerability', giving foreign computer hackers the opportunity to target individual homes, municipal buildings and even whole districts. Described by security experts as the 'modern day equivalent of a nuclear strike', hackers would be able to switch off meters from overseas, cutting off targets from the national grid.

The Observer cautioned that:

> the capacity for remote cutoff in a networked system opens up a huge national cybersecurity vulnerability. After all, if E.ON can remotely disconnect every house in East Anglia, so too can a hacker in China. Still, if that happens we'll still be able to read, by candlelight, the OED's top definition of 'smart': 'an instance of sharp physical pain, especially caused by a blow, sting or wound.'

During the same time, the energy department, UK government, and later on the national campaign of the smart meter rollout argued that privacy issues were taken seriously and were being addressed—an implicit recognition of the powerfulness of this particular dystopian imaginary.

Families in turmoil

The final negative imaginary of "families in turmoil" relates to smart meters adding stress or tension to family routines, or worse, breaking families apart.

The primary coalition here is consumer groups and statements made my parents.

For instance, *The Sunday Times* noted that one writer mentioned how smart meters provoked their seven-year-old daughter to shout: …

> "you're destroying the planet, Daddy" as she stared into the display unit of my new smart meters. It had seemed an ordinary Sunday morning until then. The tumble dryer was whirring downstairs, the kettle was boiling for my second mug of tea, and a few lights were on here and there.

The Daily Mail wrote that smart meters enabled another parent to "become the amusing nag around the house." Another newspaper wrote that one person complained their smart meter "turned the children into a kind of eco-police force." *The Independent* wrote that "in some trials, the meters … yielded negligible savings—and often at the expense of family unity, with people bickering over energy usage" and that "tension revealed in the study ranged from light-hearted to heated, and sometimes a mixture," with one married couple apparently almost "breaking up" over how a smart meter implicated boiling water in a tea kettle.

Conservatism, agency, and contestation in smart meter futures

The analysis of smart meters in this chapter draws attention to both future-orientated discourses (what is being envisioned) and where they are located (where it is said and by whom). Utopian imaginaries are mainly grounded in conservative and financially rationalized storylines whereas dystopian imaginaries were based on contesting, emotive, and issue specific storylines.

Utopian imaginaries: conservative and financially rationalized storylines

The two most prevalent utopian imaginaries of "empowered consumer" and "energy conscious world" are mainly grounded in short-term, conservative futures where the current energy system can be essentially maintained. References to the smart grid that are linked to more transformative social and technological changes are rather limited in the newspaper articles (see "low-carbon grid"). Moreover, the two prevailing utopian imaginaries appeal to rational or utilitarian calculations, making apparent the strong dominance of financial rationales referring to energy use and energy efficiency savings on bills. Potential wider societal benefits of creating sustainable, more democratic energy systems are rarely discussed in the newspapers. Further, it is mainly government actors who construct storylines of a smart grid, rather than a diverse set of actors creating multiple futures of the grid within the newspapers.

In terms of the coalitions within the utopian imaginaries, storylines morphed across institutions over time; different institutions change their stance or promote contradictory and inherently malleable futures. Some coalitions initially supported the prevalent imaginaries of "empowered consumer" and "energy conscious world" only to later abandon them. From 2012, environmental think tanks "left" those imaginaries and began to develop dystopian imaginaries, leaving government actors as the main source of advocates of the current model of the rollout.

Dystopian imaginaries: contesting, emotive, and issue specific storylines

These dystopian imaginaries were not necessarily grounded in opposing the SMIP, considering that environmental think tanks and consumer protection organizations were part of constructing prevalent utopian imaginaries between 2006 and 2012. They rather contest these utopian imaginaries, i.e. "empowered consumer" with "astronomical bills" and "energy conscious world" with "costly disaster." Contestations surrounding who benefits from the rollout (and how) were partly reduced to competing ideas of whether energy savings could be made in the home and whether technologies enabled (or not) certain kinds of technological promises. Somewhat missing from the newspaper articles were attempts to opening up the smart energy debate to broader issues of democracy and justice linked to a smart grid.

Rather than only contesting the utopian imaginaries, some dystopian imaginaries presented issue specific storylines mainly grounded in privacy and security concerns. Particularly stark storylines were obviously intended to invoke strong emotional reactions. Such storylines made frequent use of metaphors. This included, for instance, envisioning smart meter that are a "spy in the home" (*The Telegraph*), creating a "permanent window on private life" (*The Times*) and/or leaving "families vulnerable to terrorist attacks," which are described to be the "modern day equivalent of a nuclear strike" (*Mail Online*). Within these newspaper articles, smart meters were transformed into an actor who interferes with the Englishman's notion of "my home is my castle" i.e. a private space to do what one pleases, and in the process permits surveilling activities, stealing data and creating destruction. The connectivity of web-based electronic devices, as represented in the smart meter, have an invisibility and spread that "are not easily grasped without the help of metaphor ... to give people a feeling of understanding, journalists exploit analogies with familiar phenomena ... highlighting and hiding selected features of the phenomena represented."[52]

In terms of coalitions, the dystopian imaginaries in particular suggest very different institutions behind particular storylines, resembling themes of fragmentation and diversity. For instance, the "hacked and vulnerable grid" imaginary was mainly connected to groups with an expertise and interest in security (partly fragmented from the other storylines), whereas "astronomical bills," "costly disaster," and "big brother" were made up of discourse coalitions representing a diverse set of actors. Such fragmentations and diversity seems to have created a situation where the different institutions were competing over the attention of their particular concern surrounding the rollout within the newspapers. Different parliamentary committees remain part of four coalitions within three dystopian imaginaries—suggesting that the SMIP has been extensively scrutinized and contested within the UK parliament.

Conclusions

This chapter has examined the storylines and sociotechnical imaginaries of the smart meter rollout within the UK national press. An analysis of media discourses has shown how and by whom smart meters (and their rollout) are being discussed and envisioned within the UK print media over time. Across the corpus of newspaper articles surveyed, I identified a collection of nine imaginaries. Four of these—"empowered consumers," "energy conscious world," "low-carbon grid," and "future smart innovation"—depict smart meters as catalysts for a highly desirable future. Five of these—"hacked and vulnerable grid," "big brother," "costly disaster," "astronomical bills," and "families in turmoil"—frame smart meters as socially regressive or even dangerous.

The analysis reminds us that smart meters possess interpretive flexibility, they reflect contradiction and contestation, and reveal that there is no uniform imaginary of the type of grid such smart meters can deliver. Smart meters have multivalent and ambiguous meanings and technological promises. Instead of a unifying meta-discourse, I see distinct futures and storylines in tension. Part of this tension is inherent in the fragmented and issue specific storylines themselves; some imaginaries are in conflict with others, e.g. "inclusive innovation" is about benefitting industry, whereas "empowered consumers" comes partly at the expense of industry. Others are grounded in directly opposing imaginaries, e.g. "energy conscious world" is about making the use of energy more efficient through technologies, whereas "costly failure" is grounded in the technologies failing to achieve envisioned benefits. This fragmentation of imaginaries could reflect the relatively emergent nature of the technology: smart meters are fairly new devices, implying they have not yet reached the stabilization and transference phases Felt suggests more mature technologies such as biotechnology or nuclear weapons may have. Instead, smart meters appear to be more in the discordant multiple rehearsals phase. Similarly, if imaginaries are about Jasanoff's materiality, meaning, and morality, those dimensions are still being challenged in terms of the SMIP: material benefits (cost savings, emissions reductions, safety) uncertain, meaning disputed, morality questioned.

Critically, three interesting themes emerge: that of fluidity over time, and those of fragmentation and diversity. In terms of fluidity, storylines and imaginaries change and morph across institutions over time—different institutions change their stance or promote contradictory imaginaries—from supporting certain futures to openly contesting them in relation to struggles over technological promises and implementation ideas (see "empowered consumer" to "astronomical bills"). In terms of fragmentation and diversity, the dystopian smart meter imaginaries in particular suggest either different institutions or a diverse set of institutions behind particular imaginaries and often issues-specific storylines (such as privacy concerns). These issue-specific storylines connected to dystopian imaginaries seem to compete for attention, potentially struggling to gain traction beyond making people aware of particular issues.

Furthermore, the existence of utopian and dystopian imaginaries implies that the smart meter rollout is being publically (and vigorously) debated and deconstructed, reflecting multiple possible pathways and futures. In this way, media discourses can open up the politics of knowledge production to a wider audience: drawing attention to privacy concerns and potentially rising energy costs, to name a few. Then again, newspapers might also follow their own agendas surrounding the rollout through a prevalence of either dystopian or utopian imaginaries. In this sense, the media might act as far more than a process of mediation between expert knowledge and public comprehension.[53] The imaginaries themselves influence both actors and institutions, at times raising issues otherwise ignored, while at other times limiting possible futures and topics of deliberation and discussion that follow.

Notes

1   Sovacool, B. K. et al. 2017, October. Vulnerability and Resistance in the United Kingdom's Smart Meter Transition. Energy Policy 109: 767–781.
2   Lewis, D., and J. Kerr. 2014. Not Too Clever: Will Smart Meters Be the Next Government IT Disaster? (London: Institute of Directors).
3   Sovacool et al. (2017).
4   Hielscher, S., and B. K. Sovacool. 2018, September. Contested Smart and Low-Carbon Energy Futures: Media Discourses of Smart Meters in the United Kingdom. Journal of Cleaner Production 195: 978–990.
5   Marshall, J. P. 2016. Disordering Fantasies of Coal and Technology: Carbon Capture and Storage in Australia. Energy Policy 99: 288–298.
6   Eames, M. et al. 2006. Negotiating Contested Visions and Place-Specific Expectations of the Hydrogen Economy. Technology Analysis & Strategic Management 18 (3–4): 361–374.
7   McDowall, W., and M. Eames. 2006. Forecasts, Scenarios, Visions, Backcasts and Roadmaps to the Hydrogen Economy: A Review of the Hydrogen Futures Literature. Energy Policy 34 (11): 1236–1250.
8   Hultman, M. 2009. Back to the Future: The Dream of a Perpetuum Mobile in the Atomic Society and the Hydrogen Economy. Futures 41 (4): 226–233.
9   Kuchler, M. 2014. Sweet Dreams (Are Made of Cellulose): Sociotechnical Imaginaries of Second-Generation Bioenergy in the Global Debate. Ecological Economics 107: 431–437.

10 Fatimah, Y. A. 2015. Fantasy, Values, and Identity in Biofuel Innovation: Examining the Promise of Jatropha for Indonesia. Energy Research & Social Science 7: 108–116.

11 Levidow, L., and T. Papaioannou. 2013. State Imaginaries of the Public Good: Shaping UK Innovation Priorities for Bioenergy. Environmental Science & Policy 30: 36–49.

12 Cloke, J., A. Mohr, and E. Brown. 2017, September. Imagining Renewable Energy: Towards a Social Energy Systems Approach to Community Renewable Energy Projects in the Global South. Energy Research & Social Science 31: 263–272.

13 Hjerpe, M., and B.-O. Linner. 2009. Utopian and Dystopian Thought in Climate Change Science and Policy. Futures 41 (4): 234–245.

14 Litfin, K. T. 1994. Ozone Discourses: Science and Politics in Global Environmental Cooperation (New York: Columbia University Press).

15 Darby, S. 2008. Energy Feedback in Buildings: Improving the Infrastructure for Demand Reduction. Building Research & Information 36 (5): 499–508.

16 Darby, S. 2010. Smart Metering: What Potential for Householder Engagement? Building Research & Information 38 (5): 442–457.

17 Hielscher, S., and P. Kivimaa. 2018. In press. Governance through Expectations: Examining the Long-Term Policy Relevance of Smart Meters in the United Kingdom. Futures.

18 House of Commons Science and Technology Committee. 2016, September 24. Evidence Check: Smart Metering of Electricity and Gas.

19 Darby (2010).

20 Sovacool et al. (2017: 767–781).

21 Darby, S. 2009. Implementing Article 13 of the Energy Services Directive and Defining the Purpose of New Metering Infrastructures. Proceedings of the European Council for an Energy-Efficient Economy Summer Study. Paper(2262).

22 Hielscher and Sovacool (2018: 978–990).

23 [BEIS] Department for Business, Energy & Industrial Strategy. 2018, May 31. Smart Meters Implementation Programme 2018 Progress Update. www.gov.uk/govern ment/statistics/statistical-release-and-data-smart-meters-great-britain-quarter-1-2018.

24 Anderson, B. 1991 [1983]. Imagined Communities Revised Edition (London: Verso).

25 Appadurai, A. 2002. Disjuncture and Difference in the Global Cultural Economy, Theory Culture Society 1990 (7): 295–310

26 Taylor, C. 2004. Modem Social Imaginaries (Durham, NC: Duke University Press).

27 Marcus, G. E. (Ed.). 1995. Techno Scientific Imaginaries: Conversations, Profiles, and Memoirs (Chicago, IL: University of Chicago Press).

28 Sovacool, B. K., and D. J. Hess. 2017, October. Ordering Theories: Typologies and Conceptual Frameworks for Sociotechnical Change. Social Studies of Science 47 (5): 703–750.

29 Jasanoff, S. 2015a. Imagined and Invented Worlds. In Jasanoff, S. and Kim, S. H. (Eds.): Dreamscapes of Modernity: Sociotechnical Imaginaries and the Fabrication of Power (Chicago, IL: University of Chicago Press), pp. 321–343.

30 Jasanoff, S., and S. H. Kim. 2009. Containing the Atom: Sociotechnical Imaginaries and Nuclear Power in the United States and South Korea. Minerva 47 (2): 119–146.

31 Jasanoff, S. 2015b. Future Imperfect: Science, Technology, and the Imaginations of Modernity. In S. Jasanoff and S. H. Kim (Eds.): Dreamscapes of Modernity: Sociotechnical Imaginaries and the Fabrication of Power (Chicago, IL: University of Chicago Press), pp. 1–33.

32 Jasanoff (2015b: 24).

33 Jasanoff, S. 2015c. Imagined and Invented Worlds. In Jasanoff, S. and Kim, S. H. (Eds.): Dreamscapes of Modernity: Sociotechnical Imaginaries and the Fabrication of Power (Chicago, IL: University of Chicago Press), pp. 220–228.

34 Pickersgill, M. 2011. Connecting Neuroscience and Law: Anticipatory Discourse and the Role of Sociotechnical Imaginaries. New Genetics and Society 30 (1): 27–40.

35 Jasanoff, S., and S.-H. Kim. 2013. Sociotechnical Imaginaries and National Energy Policies. Science as Culture 22 (2): 189–196.
36 Jasanoff and Kim (2009: 119–146).
37 Ibid.
38 Felt, U. 2013. Keeping Technologies Out: Sociotechnical Imaginaries and the Formation of a National Technopolitical Identity; Pre-Print; Published by the Department of Social Studies.
39 MacKenzie, D., and G. Spinardi. 1995, July. Tacit Knowledge, Weapons Design, and the Uninvention of Nuclear Weapons. American Journal of Sociology 101 (1): 44–99.
40 Kivimaa, P., and F. Kern. 2016. Creative Destruction or Mere Niche Support? Innovation Policy Mixes for Sustainability Transitions. Research Policy 45 (1): 205–217.
41 Felt (2013).
42 Jasanoff (2015b: 19).
43 Jasanoff (2015c).
44 Hultman (2009).
45 Hielscher and Sovacool (2018: 978–990).
46 Porter, K. E., and M. Hulme. 2013. The Emergence of the Geoengineering Debate in the UK Print Media: A Frame Analysis. The Geographical Journal 179 (4): 342–355.
47 Dirikx, A., and D. Gelders. 2010. Ideologies Overruled? an Explorative Study of the Link between Ideology and Climate Change Reporting in the Dutch and French Newspapers. Environmental Communication 4 (2): 190–205.
48 Hielscher and Sovacool (2018: 978–990).
49 Ibid.
50 NAO. 2011. Update on Preparations for the Roll-Out of Smart Meters (London: National Audit Office).
51 PAC. 2011. Preparations for the Roll-Out of Smart Meters. Sixty-Third Report of Session 2010–2012 (London: House of Commons: Committee of Public Accounts).
52 Nerlich, B. 2007. Media, Metaphors and Modelling: How the UK Newspapers Reported the Epidemiological Modelling Controversy during the 2001 Foot and Mouth Outbreak. Science, Technology, & Human Values 32 (4): 432–457.
53 Becken, S. 2014. Oil Depletion or A Market Problem? A Framing Analysis of Peak Oil in the Economist News Magazine. Energy Research and Social Science 2: 125–134.

7

EXPECTATIONS

Electric mobility and experts in the Nordic region

The decarbonisation of energy and transport systems in particular is among one of the most important international challenges.[1,2,3,4] In this context, due to the transportation sector's dependence on fossil fuel energy sources and the monumental negative consequences in terms of climate change, air pollution and other social impacts, countless researchers, policymakers and other stakeholders view a widespread transition to electric mobility as both feasible and socially desirable.[5,6] The International Energy Agency even projects in its most recent *World Energy Outlook*, under the "Sustainable Development Scenario," that 875 million electric vehicles will need to be adopted by 2040.[7]

One potentially fruitful innovation within electric mobility has been vehicle-to-grid (V2G), which refers to efforts to link the electric power system and the transportation system in ways that can improve the sustainability and security of both.[8] A V2G configuration means that personal automobiles have the opportunity to become not only vehicles but mobile, self-contained resources that can manage power flow and displace the need for electric utility infrastructure. A transition to V2G could enable vehicles to simultaneously improve the efficiency (and profitability) of electricity grids, reduce greenhouse gas emissions from transport, accommodate low-carbon sources of energy, and reap cost savings for vehicle owners, drivers, and other users.[9,10,11,12]

In this chapter, I ask: How is V2G and electric mobility being discussed, envisioned, and promoted by experts in the Nordic region? To provide an answer, the chapter draws on 257 research interviews with experts across Denmark, Finland, Iceland, Norway, and Sweden.[13] The chapter presents and analyses eight distinct visions: four positive visions of "the rapid electric society," "ubiquitous and clean automobility," "innovation nirvana," and "energy autarky" are contrasted with four negative visions of "hacked grids," "frozen families," "broken businesses," and "captive consumers." It discusses tensions and synergies between these visions as well as places them into a typology.

In proceeding as such, for those familiar with the sociology of expectation, I explore negative promises and visions alongside positive ones. Although some previous work has investigated the specific utopian and dystopian dynamics of climate change discourses,[14] none (to my knowledge) has extended this dichotomy of positive and negative visions to the domain of electric vehicles or especially V2G. In addition, I elaborate further on the discussion of ideographs,[15,16] offering an inventory that goes well beyond that of "technological progress." Finally, whereas Van Lente,[17] Berkout[18] and Michael[19] offer varying typologies of expectations centered on distinctions such as fast vs. slow or public vs. private, I offer a typology emphasizing how expectations differ meaningfully in terms of temporality (proximal vs. distant) and pace of change (incremental vs. radical).

Conceptualizing electric vehicles and vehicle-to-grid technologies

It is perhaps easiest to conceptualize an electric vehicle (EV) compared to its conventional counterpart. Most modern automobiles employ internal combustion (IC) engines, which start quickly and provide power as soon as drivers need it. By contrast, hybrid electric vehicles (HEVs), which have seen commercial success for almost 20 years now, add a battery and electric motor to a car that uses an IC engine. HEVs use the electric motor and electronics to more efficiently operate the IC engine, which can cut fossil fuel usage, greenhouse gas emissions and air pollution. HEVs do not plug in to the electrical grid, while two types of plug-in electric vehicles (PEVs) do. First, plug-in hybrid electric vehicles (PHEVs) are capable of recharging from the electrical grid, while maintaining an IC engine that allows a flexibility to power the vehicle with fossil fuels or electricity. Second, battery electric vehicles (BEVs) draw their energy for propulsion strictly from a battery. Throughout the chapter, I use the umbrella term "EV" to refer to HEVs, PEVs, PHEVs, and BEVs as it is simpler that way.

The deployment or EVs in vehicle-to-grid (V2G) configuration relates to what they do when plugged into the home, office building, charging point, or grid. V2G is a fairly broad concept that describes efforts to intelligently link and plan vehicles with the electric grid.[20] California's Independent System Operator provides several categorizations of the potential for V2G, where systems can vary by three attributes: the direction of power flow (unidirectional or bidirectional), whether benefits to the grid are provided by one or multiple resources, and whether actors have unified or fragmented resources.[21]

V2G systems also can vary by the mechanism of engagement—that is, how are PEV owners and operators being incentivized to participate in such a system? Perhaps most obviously, time-of-use electricity pricing is available in some regions, where the price of electricity at any given time is tied to its availability, and changing prices are meant to control load across all electricity consumers.[22]

Time of use rates alone can lead to managed charging, providing incentive for EV users to charge their vehicles at times that are more efficient (or more environmentally beneficial), whether manually, by a simple timer system, or automatic controls. Similarly, time of use rates provide incentive for a user to enroll in a V2G system, allowing storage of electricity when rates are low and selling back to the grid when rates are high.

Kempton and colleagues have developed a framework to explore V2G which electric markets that have the greatest value and then develop systems to participate in those markets.[23,24,25,26,27] The authors determined that vehicles must possess three elements to operate in V2G configuration: a power connection to the electricity grid, a control and/or communication device that allows the grid operators access to the battery (typically via an aggregator), and precision metering to qualify for fast-response markets, and provide auditability.[28,29] Other phrases used to describe V2G concepts include V2X[30] (to signify vehicle-to-home, vehicle-to-building, vehicle-to-community and vehicle-to-utility configurations) as well as "mobile energy storage systems,"[31] "battery-to-grid,"[32] "gridable vehicles"[33] "virtual power plants,"[34] and "S3Ps" (small portable power plants).[35] A vehicle capable of V2G can in theory offer at least eight different types of grid services, including: active power regulation, supporting reactive power, load balancing by valley filling, current harmonics filtering, peak load shaving, reduction in utility operating cost and overall cost of service, improved load factors, and the tracking of variable renewable energy resources.[36]

This chapter focuses on expert expectations (and other thoughts) about V2G and EVs in the Nordic region, one of the first places V2G is being piloted, with Denmark a noted leader in testing and research.[37] Moreover, the Nordics are generally a comparable group of countries that show a variety of EV adoption rates, including the globally leading Norwegian EV adoption rate, increasing rates in Iceland and Sweden, the lessons from Denmark's shifting EV incentives, and Finland's struggle.[38,39] This makes the Nordic region a relevant and comparable case study with globally applicable lessons.[40]

The sociology of hype and future expectations

To explore the visions associated with electric mobility and V2G, the chapter relies on concepts arising out of the "sociology of expectation" (sometimes called "expectation studies," a "sociology of the future" or the "sociology of anticipation"). This approach aims to assess how "guiding visions" or "normative expectations" about future benefits affect and structure technology.[41,42] Bakker et al. define "technological expectations" as "real-time representations of future technological situations and capabilities. That is, it is a combination of expected progress of the technology at stake, its future markets, and its societal context."[43] Van Lente is even more precise in offering a definition: "expectations are circulating representations of the future."[44] Such expectations can be individual or collective, and they can involve statements, images, graphs, terms, and stories, within or between firms, research groups, policy, society. Concepts from the sociology of expectation attempt to reveal the "narrative infrastructure" or "mosaic of stories" surrounding technology.[45]

But how do expectations originate and circulate? A variety of concepts currently ground the approach. One is the notion of a *rhetorical vision*: advocates of a particular technology will often hold shared expectations and narratives about it. These will have specific dramatic elements such as plot lines, stories and characters. Van Lente terms this "mutual positioning:" actors position themselves, others and future technology in a story (or a plot), and so make others into characters in the story.[46] This serves a basic coordinating power that creates a shared agenda and a division of tasks. Many times these characters fall into archetypes such as "the user," "the ally," "the adversary," and the "product to be."[47] Oftentimes, visions become shared, a "collectively held and communicable schemata that represent future objectives and express the means by which these objectives will be realized."[48] Visions are most powerful when they become part of a "collective repertoire" of ideas and statements shared by large stakeholder groups; in such contexts, the vision cannot be ignored even by those that do not share its ideas, for the vision forms a part of social reality.[49] Visions at this scale become multi-scalar and multi-actor: they can serve as motivating factors for researchers and inventors, steer national policy, shape research patronage and funding patterns, and blossom through meso level social and innovation networks.[50]

To be clear, within this body of work, expectations are not synonymous with a narrative or rhetorical vision, nor with a promise. An expectation would be preparing London for electric vehicle charging or that local-level actors would resist electric taxis, whereas a vision would refer to the broader narrative story-lines revolving around themes of independence, power, justice, and sense of community.[51] Promises can also take many forms: umbrella promises tend to be vague, open ended, and non-falsifiable; more specific promises can be closed and falsifiable, and thus more prone to disappointment.

One particularly powerful type of collective vision or narrative is an *ideograph*, a sort of meta-story, meta-vision or super-promise that cuts across visions and recurs. Van Lente suggests than an ideograph intertwines ideology, power, social control and language—it becomes a "way of understanding what collective conviction means." Ideographs signify a baseline of public and political commentary, and often relate to common rhetorical tropes such as "freedom," "quality," "prosperity," and "safety." Van Lente muses that perhaps the most prominent ideograph connected to technology (in the past) has been that of "continual progress," an idea reaching as far back as the Enlightenment and one connected to sociotechnical systems such as electricity, information and communication technology, biotechnology, and nanotechnology.[52]

Another concept is the notion of a *promise—requirement cycle*. Van Lente suggests that technology does not start with problems but with promises, which can be taken up in agendas (groups, firms, policy) and lead to requirements and protection to continue.[53] Promises and expectations of emerging technologies thus become part of an agenda-setting process that germinates into a requirement for engineers and other actors, giving them a "mandate" to develop "their" technology.[54] Steering and coordination of action occurs as expectations are voiced and responded to. In this way, the functionality of the vision results in a binding promise to developers and advocates: "the freedom to explore and develop combined with a societal obligation to deliver in the end," i.e. of "promissory commitments that become part of a shared agenda and thus require action."[55] This dynamism between promises and requirements results in a "nested phenomena" graphically depicted in Figure 7.1. As the Figure illustrates, when a promise becomes accepted as part of an agenda, more detailed expectations must be circulated and adopted. A broader promise may thus lead to more specific, achievable promises.[56,57]

A fourth concept describes the types of actors involved in brokering expectations, something termed *enactors* and *selectors*.[59] Given that there is rarely a single technological solution to a pressing social (or other type of) problem, different technologies will evoke distinct reactions from stakeholder groups. An *enactor* will stress criteria that favour their particular variation—they will be more steadfast and dogmatic, less flexible, over preferred attributes and performance aspects. Conversely, a *selector* will balance different and at times competing criteria about a technology and will be inherently fluid and dynamic. As Figure 7.2 typifies, enactors focus mostly on maintaining expectations, whereas selectors focus more

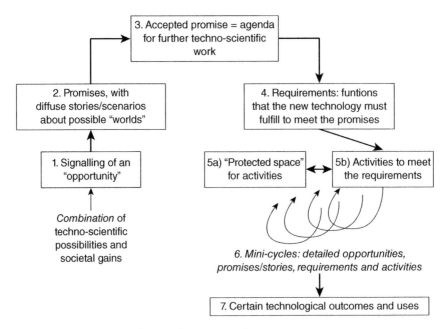

FIGURE 7.1 Dynamic evolution of promises and requirements
Source: Van Lente, H., and A. Rip 1998[58]

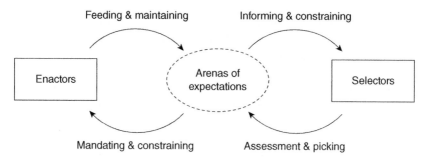

FIGURE 7.2 Enactors, selectors, and the evolutionary pressures created by arenas of expectations

Source: Bakker, S., H. V. Lente, and M. Meeus 2011. Arenas of Expectations for Hydrogen Technologies. Technological Forecasting & Social Change 78: 152–162[60]

on picking expectations. Sometimes, selectors evaluate a technology based on more objective criteria such as facts, specifications, and cost; in other cases, where technology is at earlier stages of development, more subjective criteria may be used, including only expectations. Enactors and selectors will compete for resources and attention in battlegrounds called "arenas of expectations," defined as the location where expectations are voiced and tested with experience, knowledge, and interests.

Thus, expectations are subject to "trials of strength" that also result in an ongoing process of selection and variation, resulting in a quasi-evolutionary model of technological development also characterized by Figure 7.1. Such a competitive environment—which Berkout interprets as "bids about what the future might be like, offered by agents in the context of other expectation bids"—often results in expectations that are both flexible/adaptable as well as contested.[61] This finding helps explain why many rhetorical visions are strategically contradictory: visions are malleable, allowing actors attempting to build support to avoid discussing technical details that may expose the contested nature of their own agenda.

Enactors and selectors can even come to form a distinct *technological community*,[62] defined as a group "of scientists and engineers, who are working towards solving an interrelated set of technological problems and who may be organizationally and geographically dispersed but who nevertheless communicate with each other."[63] Technological communities are not uniform; scientists in academia tend to maintain positive expectations in the face of countervailing evidence so long as they continue to receive funding or a mandate; whereas members of industry will be more skeptical and tend to compare and test different expectations.[64] Academics remain more concerned with a special field of knowledge, whereas industrial actors care more about finding solutions that work for their incumbent firms. Ultimately, more academic technological communities will emphasize a specific area that promises to be relevant for a wide audience with relatively unspecified problems, whereas industrial technological communities are concerned with the means to meet a specific goal based on the community's competencies or other grounds. In the literature this distinction has been characterized as paradigm-driven versus solution-driven communities, or between design-based versus sponsor-based communities.[65]

A final related concept is that of a *hype cycle* or *promise-disappointment cycles*, an admittedly simple but visual representation of the ups and downs of technological expectations. This helps give a common structure to the waves of hype (and disappointment) that expectations can motivate. Here technologies are seen to move along a path from a trigger to a peak in expectations, then plummeting into a trough of disillusionment before eventually giving rise to a range of somewhat more modest applications, as Figure 7.3 conveys.

To provide some empirical context, visions and hype cycles about EVs are not a new phenomenon. The promise of EVs has been touted by supporters for decades, with recurring waves of hype and disappointment. In fact, an 1899 *Scientific American* article praised electricity as "an ideal power for vehicles, for it eliminates all the complicated machinery of either gas, steam, or compressed air motors, with their attendant noise, heat, and vibration."[66] More than a century ago, the *New York Times* declared that the electric car "has long been recognized as the ideal solution" because it "is cleaner and quieter" and "much more economical."[67] Electric vehicles (EVs), in contrast to horses, bicycles, trains, steamers, and gasoline vehicles, held many benefits, and became vehicles of choice in the early 1900s. Their quieter operation enabled them to run in noise-restricted areas and more affluent neighborhoods. Many women also preferred push button

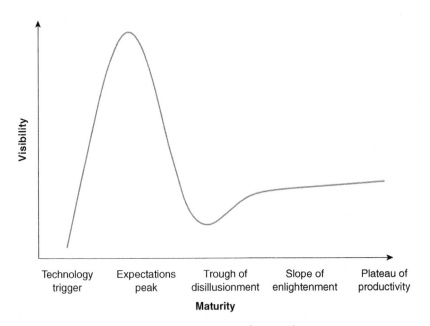

FIGURE 7.3 Visibility and maturing in expectation hype cycles

electrics as they did not require the shifting of gears or turning of hand cranks to start.[68,69,70]

In 1901, Pope's Electric Vehicle Company was both the largest manufacturer and the largest owner and operator of motor vehicles in the United States; that same year, it was also possible to travel by electric car from New York to Philadelphia, thanks to six charging stations that were built in New Jersey.[71] In 1903, Boston had 36 charging sites.[72] In Europe, electric vehicles outnumbered gasoline powered vehicles in 1900 and appeared as taxicabs, delivery vans, and even fire engines, as well as private automobiles. Commercial fleets adopted EVs as well. By 1905, Mom reminds us that "more than half of all commercial vehicles in the United States were electric powered."[73]

However, the rapid success of EVs and the resulting hype was partly responsible for their downfall. Developers and proponents of passenger and commercial EVs believed that the use of electricity as a fuel was associated with modernity and progress itself. While they were also able to cultivate an appealing vision of a horseless age, many electric vehicle proponents thus ignored some of the real advantages of gasoline and the stated preferences of customers. These advocates were so focused on an "ecstatic belief" in the future of EVs that they became blinded by the practical challenges that emerged.[74] Many were also convinced, from previous improvements in battery size and range, that a "miracle battery" would eventually be developed. In short, EV proponents believed in technological optimism and placed faith in human ingenuity to overcome lingering technical problems.

Despite these high hopes, the use of EVs slowly declined and then sharply dropped off, so much that by 1920 they constituted less than 2 percent of the overall market. Even the commercial sector slowly abandoned them: in 1913, 10 percent all commercial vehicles were electric powered but by 1925 the number had dropped to less than 3 percent.[75]

A second meaningful wave of hype has been more recent, driven by motivations about sustainability and aggressive policies in places such as Europe and California. In the late 1980s, there was a general acknowledgment among automobile manufacturers that EVs need not be confined to the narrow markets of delivery vans, golf carts, and homemade cars.[76] As one team of transport researchers predicted in the early 1990s, "By the turn of the century, electric passenger vehicles could be viable as second cars in multicar households … No longer does successful commercialization depend on technical breakthroughs."[77] General Motors similarly declared that electric propulsion for passenger vehicles was "suitable for mass production at affordable costs."[78] Substantial research programs shown in the top panel of Table 7.1 were initiated. As the IEEE wrote, "The 1990s are likely to be the decade in which the long-sought practical, economical electric vehicles will begin to be realized."[79]

However, despite such hype (and research expenditure), almost all of those EV models failed to meet sales targets, with the possible exception being Nissan's hybrid. As the bottom panel of Table 7.1 indicates, although such models had

TABLE 7.1 Research strategies, prices, and key data for electric vehicles in the 1990s

Top panel: EV research initiatives and prototypes in the 1990s at major automakers

Firm	Summary	Battery electric vehicles	Hybrid electric vehicles
General Motors	Moved away from EVs towards fuel cells	GM spent more than $1 billion developing the EV1 (incl. marketing). Only 800 vehicles were leased during a four-year period. In 1998 the last update of EV1 was launched. After relaxation of ZEV mandate (1996) GM shifted research focus towards fuel cell technology.	Until 2004 skeptical about hybrids. In 2006 research collaboration was established with DC and BMW. GM announced (in 2006) launch of full hybrid system in two SUVs (Tahoe, Yukon) for late 2007; by late 2008 they were limited available (with 25 percent better fuel economy, and $2,500 more expensive).
Toyota	Pivoted away from BEVs to HEVs	Launched electric Toyota RAV4 by 1996. Sold only a few hundred annually. Shift of R&D focus away from BEV in 1996 (relaxation of ZEV), towards HEV and fuel cells.	Pioneered with HEV in 1997 in Japan, 2000 in California. Sold around 15,000 HEVs annually in first years; by 2002 total exceeds 100,000. Estimated losses on every vehicle sold up to 2002 or 2003 of a few thousand dollars. By 2008 three models Toyota have been launched outside Japan (Prius, Highlander, Camry 2006), and three of Lexus. Total sales adding to 1.2 million.
BMW	Skeptical about EVs	No EV concept cars presented. One technical paper on R&D on E1 made public (1992).	Until 2004 skeptical about hybrids. In 2006 research collaboration was established with GM and DC. No near-future plans for full hybrid launches. They have incorporated some electronic assistance for increasing efficiency of ICE.
Honda	R&D focused primarily on gasoline and fuel cells	Launched Honda EV plus by 1996. Sold around a hundred annually. Shift of focus away from BEV in 1996 (after relaxation of	Followed Toyota with launch of Honda Insight in 1998. Honda had three hybrid models by 2000 (Accord, Civic, Insight),

(Continued)

TABLE 7.1 (Cont.)

Firm	Summary	Battery electric vehicles	Hybrid electric vehicles
		ZEV), towards HEV and fuel cells	though sales numbers are 50–80 percent lower than Toyota.
Daimler—Chrysler	Shifted to diesel and fuel cells	Launch of the Chrysler EPIC in 1997. Sold only a few hundred yearly up to 1999. After relaxation of ZEV mandate (1996) research focus shift towards fuel cell technology	Until 2004 skeptical about hybrids. In 2006 research collaboration was established with GM and BMW.Mercedes shows prototypes of S- class hybrid.
Ford	Slow follower of ICE engine innovation	Ford launched the Ranger in 1998, and sold a few hundred vehicles annually. After relaxation of ZEV mandate (1996) research focus shift already towards fuel cell technology.	Until 2004 skeptical about hybrids. Groups- member Volvo experimented with hybrid (Desiree concept in 1999) This knowledge was later used in Ford Escape (launch in 2004). Sales are under targets.
Fiat	Wary after two failed launches	Experimented with Panda electric already in 1992. Sales were very low. In 1998 Fiat tried again and launched the Seicento Electra. There were only 294 produced, unto 2002, when the production stopped.	Presented the Multipla hybrid in 2000. No launch, due to low expected sales. Reluctant on business opportunities for current launch.
Peugeot-Citroen	Disappointed with previous EV product launches	Was initially serious about EVs: by 1993 Peugeot aimed at producing 50,000 electrics in 1998. Several versions were launched. Disappointments followed when sales were only around 2,000 annually (even though these were highest in the sector).	Sceptical about full hybrids. Only prototype of Citroen Berlingo. Launch of mild hybrid: Citroen C3 with start & stop since 2005.
Volkswagen-Audi	Antagonistic towards EVs	Launch of VW CityStromer in 1995. Only 150 vehicles were sold.	Pioneered with Audi-duo in 1997 (plug-in HEV). Sold only 60 vehicles. Disappointments created skepticism on market potential of hybrids. After 2005 R&D increased (with

(Continued)

TABLE 7.1 (Cont.)

Firm	Summary	Battery electric vehicles	Hybrid electric vehicles
			Porsche) and hybrid versions of VW Touareg, Audi Q7 and Porsche Cayenne are studied but no launch date announced.
Nissan	Cautiously optimistic about hybrids	Nissan launched the electric Altra by 1998 but only sold around 50 vehicles annually.	Nissan announced an Altima hybrid for 2007, with technology bought from Toyota. With in-house technology launch announced for 2010.
Renault	Disappointed in EV sales	Was initially serious about EVs: Renault planned in 1993 to produce 4,000 electrics in 1995. Disappointment when actual sales of Clio and Express were only a few hundred annually.	Skeptical about full hybrids, and has not announced any launches.

a. Bottom panel: EV models available on the European market in 1993

Model	Seats	Range (km)	Maximum speed (km/hr)	Price of electric model	Price of gasoline model
Kewet	2	50–100	70	129,000 SEK	
Erad Junior	2	70–80	75	74,000 FRF	48,500 FRF
VW Golf	4	71	100	288,000 SEK	110,000 SEK
Microcar Lyra	2	65	75	146,000 FRF	69,900 FRF
Puli City	2	50–80	65	88,000 SEK	
Elektro Marbella	4	50–100	80	149,000 SEK	60,000 SEK
Renault Clio	4			177,000 FRF	79,500 FRF

Source: Author, adapted from Dijk, M., and M. Yarime 2010 and Cowan and Hulten 1996.[80,81] Note: Prices include batteries and are for basic models. ICE = internal combustion engine. EV = electric vehicle. ZEV = zero emissions vehicle. HEV = hybrid electric vehicle. SUV = sport-utility vehicle. BMW = Bavarian Motor Works. VW = Volkswagen. R&D = research and development. GM = General Motors. DC = Daimler-Chrysler. SEK = Swedish Krona. FRF = French Francs

considerable range (all between 50 and 100 kilometers per charge), they were still considerably more expensive than traditional cars. EVs as a whole for passenger transport had truly dismal sales and production figures, with major European manufacturers Renault, Citroen, and Peugeot all selling only 100 to 200 models, to say nothing of General Motors losing roughly $1 billion on their endeavor promoting the EV-1 in the United States.[82]

Such cycles of hype and disappointment do not exist purely in the realm of language and rhetoric; they can significantly shape policy, media attention, and innovation patterns (such as patents). Successive patterns of hype can create considerable pressure on policymakers, especially those looking for positive stories or eager to show ambition—they can thus "jump on the bandwagon" and only reinforce and inflate the expectations.[83] Melton et al. used media content analysis to literally plot corresponding hype (news coverage) of electric vehicles in the *New York Times*, results (shown in Figure 7.4) that suggest media attention mimicked the rise and fall of interest from the automakers from 1990 to 2000.[84] Lastly, the number of EV patents and new product launches in the period 1990 to 2005 clearly indicates a cycle of hype and then "doubt and disappointment." In Japan, the number of patent applications on EVs rose steeply in the 1990s, stabilized in 1995 and declined rapidly afterwards.[85] After grappling with failures in the early 1990s, in Europe around 80 percent of automotive patents were awarded to conventional internal combustion engine related technology, against only about 20 percent for technologies associated with pure battery EVs and hybrid-electric vehicles—evidence of substantial shift in research momentum.[86] Indeed, the collapse of EV markets in the mid-1990s may have even stigmatized electric mobility to the point where firms were overly reluctant to reinvest in electric propulsion as a profitable strategy. EVs remained confined to niche markets and a few large-scale demonstration projects across Europe and the United States for most of the 2000s.

Thus, at its core, the sociology of expectation (and its related concepts) offers a semiotic and symbolic understanding of diffusion and the social acceptance or rejection of technology; diffusion is intimately tied to cognitive elements such as values, attitudes and expectations, sometimes even those that are unarticulated or hidden in the subconscious. Diffusion is a symbolically interactive process as well, with the meanings attached to technology being constantly modified, malleable, and interpretive.[88] Expectations have "heterogeneous ingredients" and "prospective structures" that are highly iterative: circulating futures exert force because they allocate roles and frame actions and reactions.[89] The theory thus demonstrates how expectations are continually "constitutive" or "performative" in defining roles and in building obligations to support a particular technology.[90,91,92] In this context, expectations serve a real-time, active purpose in shaping present day arrangements.[93] Lastly, and critically, identifying competing expectations and visions can further reveal the often implicit values behind particular sociotechnical pathways, especially to the extent that certain stakeholders may benefit at the expense of others; it can thus allow for more transparent debate and socially responsible discussion.[94]

Research design and empirical strategy

To collect original data in exploring narratives and visions of electric mobility through the lens of the sociology of expectation, my primary tool was 227 semi-

(a)

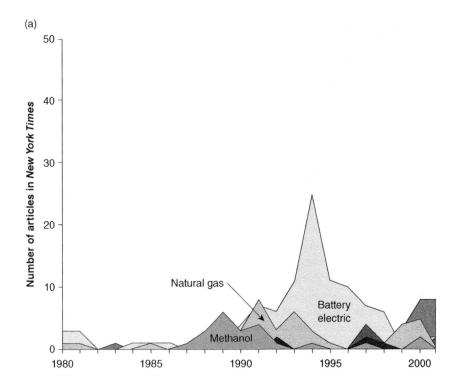

(b)

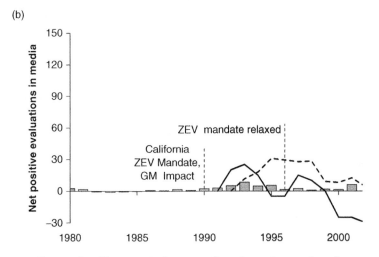

FIGURE 7.4 Expectation "hype cycles" surrounding alternative modes of transport in the *New York Times*, 1980–2000. (a) Top panel: media attention for alternative fuel vehicle technologies. (b) hype and disappointment cycle for electric vehicles

Source: Adapted from Melton, N., J. Axsen, and D. Sperling 2016.[87] Note: ZEV = zero emission vehicle. GM = General Motors

structured interviews with 257 expert participants from over 200 institutions across each of the five Nordic countries. The choice for expert (or "elite") semi-structured interviews follows the complexity of the topic of electric mobility and V2G, as semi-structured interviews allow for a timely and in-depth discussion of topics and where values play an important role. Table 7.2 offers an overview of my interviews and respondents by country, gender, focus area, and sector.

Those interviewed in Denmark, Finland, Iceland, Norway and Sweden were selected to represent the diverse array of stakeholders involved with electric mobility, from both a transport and an electricity side, and include:

TABLE 7.2 Overview of Nordic electric mobility research interviews and respondents

	Interviews (n=227)	Respondents (n=257)	Percent of Respondents
Country = Iceland (September–October 2016)	29	36	14.0
Country = Sweden (November–December 2016)	42	44	17.1
Country = Denmark (January–March 2017)	45	53	20.6
Country = Finland (Mar 2017)	50	57	22.2
Country = Norway (April–May 2017)	61	67	26.1
Gender = Male	186	215	83.7
Gender = Female	49	58	22.6
Focus[a] = Transport or Logistics	73	81	31.5
Focus = Energy or Electricity System	63	75	29.2
Focus = Funding or Investment	10	12	4.7
Focus = Environment or Climate Change	12	16	6.2
Focus = Fuel Consumption and Technology	22	23	8.9
Focus = Other	13	14	5.4
Focus = EVs and Charging Technology	34	36	14.0
Sector[b] = Commercial	66	70	27.2
Sector = Public	37	46	17.9
Sector = Semi-Public	40	51	19.8
Sector = Research	37	39	15.2
Sector = Non-Profit and Media	12	13	5.1
Sector = Lobby	22	25	9.7
Sector = Consultancy	10	10	3.9

a Focus represents the primary focus area of the organization or person in question.
b Sector represents the sector the company was working in (semi-public referring to commercial companies owned by public authorities, like DSOs).

- National government bodies, including the Ministry of Industries & Innovation (Iceland), Ministry of Environment and Energy (Sweden), Ministry of Finance (Finland), and Ministry of Taxation (Denmark):
- Local government ministries, agencies, and departments including the Akureyri Municipality (Iceland), City of Stockholm (Sweden), Aarhus Kommune (Denmark), City of Tampere (Finland), City of Oslo (Norway), and Trondheim Kommune (Norway);
- Regulatory authorities and bodies including the National Energy Authority (Iceland), Danish Transport Authority, Helsinki Regional Transport Authority (Finland) and Trafi (Finland);
- Universities and research institutes including the University of Iceland, Swedish Environmental Institute, DTU (Denmark), Aalborg University (Denmark), VTT Technical Research Centre (Finland), NTNU (Norway), and the Arctic University of Norway;
- Electricity industry players such as ON Energy (Iceland), E.ON (Sweden), Vattenfall (Sweden), Energinet (Denmark), DONG (Denmark), Fingrid (Finland), Elenia (Finland) and Statnett (Norway);
- Automobile manufacturers and dealerships including the BMW Group (Norway), Volvo (Sweden), Nissan Nordic (Finland), Volkswagen (Norway), and Renault (Denmark);
- Private sector companies including Siemens Mobility (Denmark), Nuvve (Denmark), Fortum (Finland), Virta (Finland), Clever (Sweden), Nordpool, (Sweden), Norske Hydrogen (Norway), Microsoft (Norway) and Schneider Electric (Norway);
- Industry groups and civil society organizations such as Danske Elbil Alliance (Denmark), Finnish Petroleum and Biofuels Association, Tesla Club (Finland), Power Circle (Sweden) and the Norwegian Electric Vehicle Association.

On average, the interviews lasted between thirty and ninety minutes and the interviewees were asked, among other questions: "What do you see as some of the most significant benefits, and barriers, facing EVs and V2G?" Each interview was recorded and then fully transcribed, and each participant was given a unique respondent number (e.g., R257).

Results: eight contested sociotechnical visions and promises

As this section of the article demonstrates, my interview material led to no shortage of different visions. Here, I extrapolate on the eight that were the most frequently mentioned. Table 7.3 offers an overview of how they differ by type, promises and requirements, and ideographs. After positioning them into two classes—positive and negative—I roughly ordered them by their timeframe (proximal and near-term to distant and far-term).

TABLE 7.3 Summary of expert visions, promises, and ideographs with Nordic electric mobility

Vision	Description	Promises and requirements	Ideograph(s)	Features	
				Time	Change
Rapid electric society	Electricity will come to meet all passenger transport needs or even all transport needs	Rapid charging, electric highways, adequate vehicle range, electrification	Progress, convenience	Proximal	Incremental
Ubiquitous and clean automobility	Automobility will expand indoors to include mobility within buildings	Zero carbon mobility, zero emissions vehicles, avoidance of inclement weather	Environmental sustainability, physical shelter	Proximal to distant	Radical
Innovation nirvana	Electric mobility will catalyze a cascade of continuing innovations leading to further technical breakthroughs and progress	Bigger batteries, automated vehicles, flying vehicles, robot assisted mobility, hydrogen fuel cells, coupling of innovations	Progress	Distant	Radical
Energy autarky	A transition to electric mobility will include a transformation to decentralized, local sources of energy	Self-sufficiency, community ownership, independence from energy companies, free energy	Liberty, autonomy	Distant	Incremental
Hacked and vulnerable grids	A vehicle-grid-integrated, interconnected economy would raise serious risks of loss of privacy, terrorism, and the collapse of local grids	Monitoring and surveillance, data breaches, terrorist attacks, blackouts	Privacy, security	Proximal	Radical

(Continued)

TABLE 7.3 (Cont.)

Vision	Description	Promises and requirements	Ideograph(s)	Features	
				Time	Change
Frozen families	EVs will breakdown during snowstorms, on mountains, or during emergencies, and lack sex appeal	Stranded vehicles, traffic accidents, sexless marriages	Safety, love	Proximal	Incremental
Broken and bankrupt businesses	EVs are an inferior product that will be perpetually confined to small niche markets	Bankruptcy of companies, collapse of EV markets	Employment, economic growth	Proximal to distant	Incremental
Captive consumers	EVs will only create or deepen dependences on electricity suppliers, charging companies, or battery manufacturers	Unfair tariffs, massive profits for companies	Liberty	Distant	Incremental

Source: Author, based partly on Sovacool, B. K., J. Kester, L. Noel, and G. Zarazua de Rubens in press 2019[95]

The rapid electric society

By far the most prevalent vision expressed in my interviews—across the entire sample of statements—was that of the rapid electric society. This vision merges together various promises and expectations (rapid charging, electric highways, adequate range) with a vision of a fast transformation to reliance on electricity for mobility. Sometimes, this shift to electricity is framed narrowly for passenger cars, at other times it encompasses the complete penetration of electricity for all transport modes and markets, or even new spaces and applications. Because of its proximal nature, many of its claims are more specific and falsifiable than other visions. This vision connects most frequently to ideographs such as progress and convenience.

What is particularly striking about this vision is that while it prophesizes far-reaching changes, it suggests these will be rapid and almost inevitable. For instance, as R4 states:

> It is no question that in Iceland, every car—yes, every single car—will be fueled by electricity in a few years. There is no question about it. It is obvious. We have plenty of everything to create electricity. I say there is no problem to build charging stations to be able to cover all transport needs, everything.

R16 was quick to clarify that this was not confined to Iceland or even the Nordic region. According to them:

> In 2020, passenger cars around the world should be able to drive the first 40 kilometers on something other than fossil fuels.

Similarly, R37 stated that:

> I believe a complete transition to full battery electric vehicles will take five years. After five years from now, every manufacturer will offer their own electric cars. A customer will not be able to buy anything else—there will be sleek luxurious models by Mercedes, fast ones by BMW, efficient ones by Volkswagen. Already, the big manufacturers, they're coming out with new models every year or two, with real electric cars where you can drive three hundred, four hundred kilometers on a single charge.

These statements—and no less than forty others within the sample—all portrayed a shift to electric mobility as desirable and fast, occurring in the next two to fifteen years.

In articulating the strength and veracity of this vision, some respondents were quick to use sounds, clever anecdotes, and even colorful language. R1 indicated that:

> Adoption will be so quick, it will be as if they were injected with a virus and like zombies go and buy electrical cars.

R9 discussed the likelihood of positive social feedback loops accelerating EV uptake:

> To me, the vehicle market is like in the fields of Africa. Imagine the Serengeti, you have 1 million calves all eating grass [makes mooing sound]. And then four lions approach. They see this little calf, and they would like to eat the calf, its weak, it's the easiest prey, so they run towards the calf. 10 of the calves see the lions and start running, and then 10 others think, ah ha, others are running, so they start running, and all of a sudden the whole herd is running. They come to a river and they cross the river, and

crocodiles are there grabbing you but the herd goes over the river, they make it to the other side and start eating the grass. Now, the point of the story is it took only four lions to move a million calves across a field and over a crocodile infested river to eat grass they could have eaten to begin with. Electric vehicles are those lions on the vehicle market, and they will push a herd mentality with the same force and velocity.

In generating more complex and falsifiable expectations, other respondents discussed the sorts of infrastructural shifts that would occur as society become fully electric and vehicle-to-grid capable. R83 noted that:

Not so far in the future, we will reach an almost fully electric society. For cars, we can do maybe thirty or forty kilowatts up and down without damaging the battery, because the chemistry just gets better and better … We are already seeing rapid improvements in battery technology every six months. There's huge potential for doing all sorts of grid services. EVs will be on every road, in every parking lot. Taxis, trains, and buses will all be electric. If someone, like a CEO, arrives in two or three years in the airport in Copenhagen, he walks out in the street and sees it's dirty like ten years ago, all the diesel taxis are standing idling and all that. And you fly to Norway, Sweden or Amsterdam or whatever and see everything is electric, it's quiet, and it's nice. He's going turn his heels and walk straight back to the Copenhagen airport and fly somewhere else and start his business. The future of business is electric.

R148 emphasized the sheer amount of happiness such a transition would bring to drivers and passengers:

EVs are so safe, reliable, and quiet. From our perspective, our Leaf drivers in the Nordics are our most satisfied customers in the history of our company. I mean, any product any market ever, these are our happiest people. They love the product we are selling and write to us not with complaints but with praise.

A final twist on this narrative was bringing electricity through EVs into other areas of society. Interestingly, and creatively, some respondents mused on the potentially fantastic and unusual forms of automobility and customization that such innovation could result in. R85 suggested that:

In the future, with electric mobility well established, we will have cars that contain multiple power outlets, where you can not only charge your phone but do crazy things like have a mobile outdoor stereo system, or use it for different equipment such as motor saws in the woods, or massive torches to light road work, or even to cook food in a mobile kitchen in crisis conditions … I take all of these astronomy photos so I need power all the time I am out, so in the future I will have a big battery in my trunk that will enable vehicle-to-telescope, V2T [laughter] … Just imagine what you can use it for.

R97 added that:

Future electric cars will have incredibly useful, but customized, applications that enable the vehicle to load equipment or use certain tools. We already know of a pilot vehicle in Copenhagen that can paint posters onto walls as it drives along. The idea is that with an EV, you now have access to electric power in places you didn't before. For charging phones, or even for something like electric barbeques.

Ubiquitous and clean automobility

This vision depicts EVs and V2G systems as both clean and emissions free, as well as a necessary step towards an expansion of automobility to encompass new services and business opportunities. Because EVs do not combust fuel and have no direct tailpipe emissions, they are suitable for use indoors—within apartment buildings, schools, libraries, offices, and so on. The timing of this vision varies from proximal to distant. This vision connects most prominently to ideographs of environmental sustainability and physical shelter.

The first strand of this vision emphasizes the cleanliness of electric vehicles, the imperative of climate targets, and the perils of fossil fuels. R52 commented that:

Thanks to the high shares of renewable energy in the Nordic electricity mix, we know that this bus is being fueled with wind power. We know that it is part of climate neutral transportation and we like it because of that. Knowing that gives me a green feeling. It's as clean as it can be.

R239 stated that:

EVs are a remarkable technology when you think of it. When you get into an electric car, the technology itself is very good, it is comfortable, it is cutting emissions, you have a better consciousness when driving. So, it feels better. I think if you have these cars that go longer distances you can also cut some emissions from flying. So we are heading towards a carbon neutral society.

R34 embraced the language of technical progress and perfection as well as the futility of fossil-fuel based transport when noting that:

The era of the diesel engine, gasoline, methane, and even hydrogen is over. All roads lead to electric. Very soon, people will say, how stupid were we, bringing diesel and gasoline to the table. This was is so stupid. Methane and hydrogen—no and no! Electricity—now that is perfect. Why at this stage would you continue to invest in the diesel and combustion engine? It would be the stupidest thing to do.

R68 was even pithier in their declaration:

Personally I'm absolutely convinced that what we are seeing is the death of chemical fuels as a carrier of energy, and I hope to live long enough for all electric airplanes.

The substitution of electricity would not only enable a carbon neutral society; it could also lead to an expansion of automobility into previously unthinkable dimensions. As R43 explained:

Certain vehicles which have to be used indoors, especially industrial vehicles, are perfectly suited to electrification. I've been in a company in Uppsala, where our business idea was developing electric vehicles to meet special needs: fork lifts, indoor golf-carts, scooters, where you want to keep the indoor climate, the environment, clean. Another growing market is electric lawn mowers, which can trim indoor gardens. The environmental footprint on the climate it would be hardly measurable.

R248 was even more precise in their predictions:

In 2025 there will only be electric buses in Tromsø, and they will go every-where—outdoors and indoors. Entirely new blocks of flats and buildings will be erected with this in mind. We will have bus stations in the garage, like indoors. So it's just as easy to use the bus as your own car in the garage. Difficult to go driving when its snows, but you won't have to worry about that. Since it doesn't have emissions you have the possibilities to drive literally from desk to desk. You can park it inside, no need to search for a parking space in the ice. Just think—catching a bus without having to worry about the weather.

Innovation nirvana

A more distant and ambiguous vision of an "innovation nirvana" describes a world of rapid and sequential innovation processes, a nirvana for commercializa-tion and the continuous improvement of products.[96] When applied specifically to mobility, this vision intertwines automation and self-driving vehicles, robots, flying cars, and hydrogen innovation systems. Essentially, innovation occurs within electric vehicles but it also cascades to other segments of society. This vision most strongly connects with the ideograph of progress.

One strand of the innovation nirvana vision focuses on the narrative of improvements to electric mobility technologies. As R117 commented:

> I am convinced that EV technology is going to be superior both to conventional cars as well as whatever EV technology we have today. We are going to see continuous development and improvement like cellular phones. In five years, people are going to change the cars not because they have to get rid of the old car but because they want a new smarter car. And then the new car is going to be much smarter, and innovation will become perpetual.

R191 expected that:

> The issue of batteries will become less and less important. Soon, we will be able to go seven hundred kilometers on a single charge. One charge, and you're halfway around the world.

Another narrative within the innovation nirvana was that innovations in EVs will create knowledge spillovers into other domains, notably automated vehicles and new business models. As R34 suggested:

> Everyone will be driving electric cars within ten years. And by then, you know, probably half of the fleet will be fully autonomous. So, you won't own your car, that's a change we are going to see. But it will happen very, very fast.

R107 remarked that:

> Soon, there will be different technologies that will enable us to spend more time in our car and enhance the driving experience. This may even be not driving—there will be autonomous driving, blind people can drive themselves. Just imagine you put your children in in the car and send them away. The pathway that begins with EVs ends with robot cars and autonomous cars. I see a future for them. In ten years. We have some pilots today, in certain restricted areas, but in 10 years' time they will be everywhere.

I would empathize at this point that not everyone was sanguine about automation. R6 admitted that:

> My Tesla has an autopilot, so we drove together from the airport and back with the autopilot on, it worked and I'm shit scared of it.

Others connected EV innovation with flying cars, and hydrogen fuel cells with more distant, less falsifiable and more general claims. According to R13:

> By 2050 you have 1.2 percent oil and 40–50 percent electric and 40–50 percent hydrogen. This is what the future will look like, it just depends on how marketing and infrastructure will be built.

R34 added that:

> Autonomous cars are coming, and there will be a few other changes that will rock people's worlds, like flying cars. And that's not a joke. I was talking to someone that was on a secret project with NASA, who showed me, on their iPhone, what they were doing. They are thinking really outside of the box … The future is going to be full of exciting, unforeseen surprises.

R230 supposed that:

> I'm more excited about electric planes than cars. Your children will fly, not drive, or drive something that will fly too, to their houses. So transport right there in the air. There's a lot of space. More roads will appear in the sky.

Energy autarky

A more distant energy autarky vision is meant to convey an energy (and mobility) future dominated by local, self-sufficient sustainable energy production free from the forces of corporate capitalism, where individuals come to enhance their autonomy and ownership via decentralized community sources of energy.[97,98] This vision thus connects with ideographs of liberty and autonomy.

R144 astutely captured the essence of this vision when noting that:

> In the long-term, the vision is to create a local energy and mobility economy free from outside interference, independent from energy conglomerates. Now you can call it a nightmare or a dream, but depending on who's looking at the picture, I think the grid companies are really scared about that. They have a lot of money invested into the grid, and if you're starting to produce locally and store electricity locally that will reduce the transmissions in the grid. You're eroding their sunk investments and operating against their natural monopolies. You're rejecting both their control and their entire business mentality, which is to make money from you.

R191 went so far as to frame a V2G society as one resulting in the "death" of traditional energy companies:

There's high probability that Europe is going renewable and when coupled with grid-integrated vehicles, energy will become so easy to produce it will have very little value. There is a point where energy or electricity could even be free, or almost free, like the internet. A V2G society's essentially spelling the death of traditional fossil fuel and energy companies ... For the vast majority of the lifetime of a solar plant, your marginal cost is almost zero. You need some goats to take the grass, and you need a dude with a hose to clean it every now and then, and that's it. The marginal cost goes down, it outcompetes fossil fuels because they have to pay for the fuel, and that's why I believe in the future you can envision a scenario where energy is cheap and capacity is expensive. V2G turns upside down the whole business model of the industry, because it's democratic and decentralized.

Hacked and vulnerable grids

Rather than underscoring the positive value of interconnected and potentially autonomous and ubiquitous electric vehicles, this proximal vision sees them as a threat to identity and cyber security, similarly touching in ideographs of privacy and security.

R57 captured the thrust of this vision by noting that the linking of ICT and vehicles enabled by a V2G transition could completely transform how people are connected to the web:

> I have major concerns over privacy and how future digital citizens will interact in a V2G society. I imagine a world where companies can monitor and track your every move. Everyone could see where you are and stuff like that, if you're a one-person household people could see that the car is away. They could come into your house. It's a problem—and it opens up intimately private lives not only to companies but to others who could use or misuse the system as mass surveillance.

R62 emphasized such a system could have major security risks such as terrorism:

> To me the connection between V2G and the increased potential for terrorism is not that far off. Tomorrow's terrorism is probably completely different from yesterdays and we need to think about that when we are working with these systems. Imagine a terrorist taking control of people's vehicles or trucks. Would it be possible to get into these computers and make the cars do things they shouldn't do? Yeah. How will cyber security be protected or maintained in a world full of future terrorists and hackers? If anyone can hack the computers and find out which car I'm driving and what stores I'm shopping at and what kind of products I'm shopping for

and what kind of advertisement should be sent to my home, whether I like it or not, then the privacy of people is severely infringed … Such a system may give many people the opportunity to do the evil things.

A related narrative stressed the unintentional side effects of increased electrification straining local grids. R119 in Denmark stated that:

The big problem is supporting local grids so they can handle smarter, renewable, more flexible supply. I spoke with Better Place, a company that tried to provide battery swapping and charging infrastructure for electric vehicles. Even in their parking lot, they had to do something intelligent to avoid overloading the local grid. And that's just for 20 or 30 cars. Our national grids are state of the art, and our sources of supply are getting cleaner, but local grids are lagging behind.

R149 in Finland characterized it this way:

It's the last five meter challenge. We've been focusing so much on decarbonizing electricity supply, and large grids and power pools, less attention has been paid to local grids and distribution networks. There are stories circulating about people disrupting the entire village or town grid because they tried to charge an electric vehicle. That is a monumental challenge for local networks.

R181 noted that:

EVs could place pressure on and even risk collapsing local grids. We have these transmission and distribution hotspots that might be overloaded, because you might end up adding big peaks of consumption locally. EVs also illustrate the single point of failure issue. If I have a problem with the electricity system, then even the transportation system becomes frozen. Interconnection becomes a liability not an asset.

Frozen families

This proximal vision centers on ideographs of family safety and love, and it emphasizes the expectation that EVs will merely break down during cold weather or during emergencies, and/or will complicate family relations (freezing out men from sex, turning women cold).

R1 stated this issue of unreliability clearly when suggesting that:

When the temperature is cold, and the battery is depleted, my big worry is that I cannot go pick up the baby downtown without waiting hours for a charge.

R18 put this in the context of an emergency:

> Imagine that you're sitting at your house, and you've just gotten in for the evening, and you're EV is not charged. It's cold outside, and you have maybe 20 kilometers of range left. All of a sudden, a volcano erupts and you have to evacuate. What do you do? You're stranded, and likely dead [laughs].

R225 framed this reliability issue in terms of being stranded on a mountaintop:

> When I first learned about Tesla, my reaction was that its vehicles should be banned from mountains. When the weather is bad and you're driving up a queue in a mountain, with 20 cars in a row and your blinkers on, a Tesla might just stop in the middle of the road. It would be stranded there on the mountain, and it would be very hard to save the people inside.

A separate variant of this theme touched on family solidarity and, well, sex, or lack of it. R88 suggested that EVs could freeze or at least change the sex lives of individuals because:

> EVs are completely lacking in sex appeal. They are a car for your mother-in-law, not one you take to the beach with a blonde.

R191 clarified this further:

> Let's be honest, though. Other than Tesla, which has some power and acceleration, most EVs are downright ugly. Most of the models I see here look like a fucking dustbin. It looks like a plastic can on wheels! Where is the dick factor on that one, you know? I mean, if you put your ass in a Buddy, you immediately look eighty years old. I'd be relegated to a sexless life for sure if I bought one, forget about it!

Broken and bankrupt businesses

This proximal to distant vision frames advocates of electric mobility as charlatans pushing an inferior product doomed to fail, one that will result in failed investments and broken and bankrupt businesses. Ideographs here revolve around employment and economic growth.

This narrative often begins by pointing to the inferior attributes of electric vehicles compared to their counterparts. As R8 notes:

> People here in Iceland really like their big jeeps and the idea of freedom that they represent, that you can go up to the glacier whenever you want

to, even though what you end up doing is going to the bakery. I don't see EVs substituting for this image anytime soon.

R143 deepens the argument by arguing:

People fear EVs. They are stuck with the idea that EVs are somehow an improper, lighter semi-car, not a proper car like a gasoline version.

The seriousness of such perceived inferiority was prevalent in some statements about safety, notably that EVs would (rather spontaneously) catch fire. According to R19:

EVs scare me. When you increase the power in the battery, in such a large scale, you start to worry about the sheer amount of power in the back of the car. So even it could explode with much more force than gasoline.

This was not an isolated remark, and also came up with R212:

I will admit that I know many people who are afraid that EVs in the garage will simply catch fire, because they have read somewhere that if you don't charge correctly it can overheat and combust. Similar to the lithium ion batteries in the Samsung phones or on the Boeing Dreamliner.

For reasons such as these, many respondents constructed a narrative around EV promotion only leading to disappointment and (eventually) broken companies and stranded assets. As R32 stated:

EVs just don't make business sense. If somebody told me they were starting a business to sell only EVs in the near-future, I would close it down immediately [laughs]. Because it would be wiser for me to close than to suffer for 2 years. An EV pathway is the same as misery.

R13 added that:

The sales of battery electric cars is a huge disappointment in the world. Only Iceland and Norway are great markets percentage wise. California is not doing as expected, the same for Japan and Germany. Sales are dropping in Demark as they took out some of the tax incentives. So, in general, battery electric cars are not selling, and that is a very sad story because it is a good technology. But the market is doomed to low volumes and eventual collapse.

Interestingly, some respondents were convinced that the addition of V2G capability does not overcome the business case against EVs. As R49 noted:

For the general public electric cars are death. People want a normal car with a towing hitch or a big truck. V2G doesn't change this, it is nothing more than an unreachable dream.

R113 furthered that:

> Vehicle-to-grid is a wet dream in someone's silly mind when they saw a lot of batteries out there. It doesn't make business sense.

Captive consumers

Counter to the energy autarky vision, this final distant vision one sees an EV and V2G society as one that has merely swapped one set of corporate overlords (big automakers, oil providers) with another (energy companies, digital and ICT companies). The central ideograph here is one of liberty.

R29 illustrated part of the logic of this vision when noting:

> An electric mobility future all sounds great until you ask the question of who will control the batteries. Because people are afraid that the batteries will not last long enough, and it is very costly to get new ones. You become dependent on battery providers. And once that happens, they can charge extreme prices and reap extreme profits.

R188 added that:

> I cannot even imagine the scale of investment and effort needed to place sufficient public chargers or fast chargers to cover the eventual need for EVs. People will not really want to become dependent on some distant infrastructure for their daily travel.

R20 even framed the vision in terms of substituting oil companies with lithium companies:

> As far as I know, most of the material for batteries, especially lithium, comes from Bolivia or Columbia. Is it really better to be dependent on Bolivia and Columbia on lithium than Saudi Arabia on oil? So what is the difference? Whether it's a lithium mine in the desert of Bolivia, or an oil well in Riyadh, you have to destroy many things to get either one.

Discussion: tensions, synergies, and typologies in sociotechnical visions

As predicted by the sociology of expectations, I see considerable contestation and tensions within my sample of visions. Proximal visions such as the rapid electric society and hacked grids are presented as inevitable and fast, yet ostensibly would still need policy and financial support to occur. Ubiquitous automobility sees automobility go inward, inside buildings, whereas innovation nirvana sees it go outward, to flying cars and automated long distance mobility. The innovation nirvana sees electric mobility as a lucrative source of knowledge and profits, whereas the energy autarky vision sees the collapse of the corporate entities that would be reaping those profits, and the broken and bankrupt businesses vision is premised entirely on the financial unsuitability of EV business models. Energy autarky and captive consumers are literally opposite visions. The ubiquitous automobility vision includes narratives about how healthy and happy EV drivers are, contrasted with the insecure, hacked, frozen and broken actors depicted in all four of the negative visions.

The synergies between visions are not entirely negative. Positive synergies exist as well. For instance, an innovation nirvana would only further the cleanliness and ubiquity of mobility, and also likely increase demand for a rapid electric society. The inverse holds true as well: rapid electrification of society would likely lead invariably to new innovations and breakthroughs. The vulnerability portrayed in a hacked and vulnerable grid would only hasten the factors behind the vision of broken and bankrupt businesses.

The visions also differ meaningfully in terms of some of their constitutive elements. Here, two sets of factors seem especially meaningful. The first set refers to whether the scope of sociotechnical change brought about by the vision is incremental, pragmatic, or conventional, or instead radical, substantive, and utopian.[99] Incremental visions essentially see the future pretty much as similar to the present, taking fundamental or foundational conditions of now as the basis of their foresight. This contrasts with visions that are more progressive, substantive, or ends-oriented, in which society may differ in fundamental ways from how it exists now. When applied to EVs and V2G specifically, a radical vision would be one that depicts completely new forms of mobility such as robots, or flying cars, whereas an incremental vision portrays mobility much as we know it, with human drivers and normal vehicles. In this light, the rapid electric society, energy autarky, frozen families, broken businesses, and captive consumers are all incremental visions; ubiquitous automobility, innovation nirvana, and hacked and vulnerable grids (with new forms of terrorism) are radical.

Second is whether visions are proximal vs. distant. Distant visions are far into the future—usually at least a decade away, possibly a century away. More immediate or proximal visions would occur in a few weeks to a few years' time. Here, the rapid electric society, hacked grids, frozen families, and captive

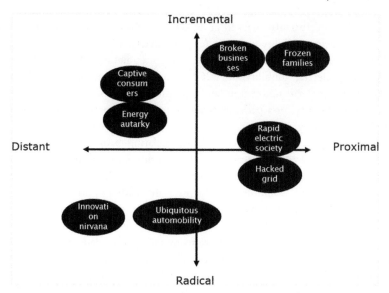

FIGURE 7.5 A typology of sociotechnical expectations for Nordic electric mobility
Source: Author

consumers would be fairly proximal; ubiquitous automobility and broken businesses more intermediate; and innovation nirvana, energy autarky, and captive consumers more distant.

Figure 7.5 maps all eight of my visions on a typology across these two dimensions. As it indicates, some visions are even twinned or interconnected. The extent that one sees more rapid electrification and the pursuit of the electrical society, the more that concerns about grid stability and hacking arise. The more that families resist EVs, fear being stranded, or express wariness about lost sex appeal, the more the broken business vision resonates, as EVs are being rejected by consumers. The more people come to adopt batteries as a means of energy autarky, the more they come to depend on charging companies, electricity suppliers, and battery manufacturers, becoming captive in another way.

Conclusion

To conclude, within the Nordic region, there remains no consensus, no master vision or ideograph, about electric mobility or V2G, what it can do, whether it is positive or negative, or how quickly it will occur. That said, some visions recurred frequently within my interview sample and were mentioned by dozens and dozens of respondents (e.g. rapid electric society, innovation nirvana, ubiquitous automobility, frozen families) whereas others were mentioned by only a handful (energy autarky, hacked and vulnerable grids, captive consumers). Some visions seemed rooted in V2G in particular (hacked grids, captive consumers) whereas others were more about electric mobility generally (electric society, broken businesses).

The imagined futures are contested and contradictory. A host of positive visions frame electric mobility or V2G systems as harbingers of an electrified society, a nirvana for innovation and technical development, a platform for automobility ubiquity, or a pathway towards energy decentralization, community control, and autarky. These starkly contrast with negative visions of families literally freezing to death, small businesses declaring bankruptcy, terrorists and hackers launching new sophisticated attacks on grids, and consumers held hostage to the whims of unsentimental corporate firms. This belies that the low-carbon transport future itself is simultaneously pregnant with opportunity and full of promise—with exciting visions I never would have predicted before undertaking my research—but also dark, despairing, and despondent. This suggests that the broader low-carbon transport future remains an open ended idea, one able to attract at least eight very different rhetorical visions.

Because of this contestation, the more specific future of electric mobility and V2G remains uncertain. Consensus at this stage across enactors and selectors should neither be expected nor sought. One person's utopia is another's dystopia. The arena of expectations reflects competing visions underpinned by competing interests and clashing values, creating intense selection pressures which perhaps lead to more pronounced variation among the visions. Put another way, a "master vision" does not exist.

Theoretically, although I confirm the utility of concepts such as rhetorical visions, ideographs, promise-requirement cycles, and enactors/selectors, the four dystopian futures seem to go against the core point of the sociology of expectations, namely this focus on the promise (instead of the problem). All the dystopian futures start from the problem and they stay there—they do not offer the relief of a promise or salvation, and in some cases sit diametrically opposed to a particular expectation or promise—i.e. vehicles breaking down, grids being hacked, businesses going bankrupt, families falling apart.

Despite this weakness, with a "sociology of expectation" lens it becomes possible to understand that the rhetoric involved in electric mobility transitions is not intrinsically tied to energy or mobility itself but rather to different social and cultural functions that electricity and mobility fulfill. In other words, the visions surrounding energy systems serve a social and psychological purpose. Visions at one narrow level about EVs and V2G end up touching on deeper ideographs of progress, convenience, environmental sustainability, physical shelter, liberty, autonomy, privacy, security, safety, love, employment, and economic growth. In this way, expectations fulfill a deeper need, and as such will likely continue to exist even as the specific sociotechnical systems behind electricity and mobility, and the distinct narratives attached to them, evolve.

Notes

1 Geels, F. W. et al. 2017, September 22. Sociotechnical Transitions for Deep Decarbonisation. Science 357 (6357): 1242–1244.

2 Figueres, C. et al. 2017. Three Years to Safeguard Our Climate. Nature 546: 593–595.

3 Eyre, N. et al. 2018. Reaching a 1.5C Target: Socio-Technical Challenges for a Rapid Transition to Low Carbon Electricity Systems. Philosophical Transactions of the Royal Society A 376 (2119).

4 Rockstrom, J. et al. 2017. A Roadmap for Rapid Decarbonisation. Science 355: 1269–1271.

5 Tran, M. et al. 2012. Realizing the Electric-Vehicle Revolution. Nature Climate Change 2: 328–333.

6 Mitchell, W. J., C. E. Borroni-Bird, and L. D. Burns. 2010. Reinventing the Automobile Personal Urban Mobility for the 21st Century (Cambridge, MA: MIT Press).

7 IEA. 2017. World Energy Outlook 2017 (Paris: OECD).

8 Sovacool, B. K., J. Axsen, and W. Kempton. 2017, October. The Future Promise of Vehicle-to-Grid (V2G) Integration: A Sociotechnical Review and Research Agenda. Annual Review of Environment and Resources 42: 377–406.

9 Kintner-Meyer, M., K. Schneider, and R. Pratt. 2007. Impacts Assessment of Plug-In Hybrid Vehicles on Electric Utilities and Regional U.S. Power Grids Part 1: Technical Analysis. Pacific Northwest National Laboratory Report. www.pnl.gov/energy/eed/etd/pdfs/phev_feasibility_analysis_combined.pdf.

10 Pasaoglu, G. et al. 2014, September. Travel Patterns and the Potential Use of Electric Cars—Results from a Direct Survey in Six European Countries. Technological Forecasting & Social Change 87: 51–59.

11 Hidrue, M. K., and G. R. Parsons. 2015. Is There a Near-Term Market for Vehicle-to-Grid Electric Vehicles? Applied Energy 151: 67–76.

12 Sovacool, B. K. et al. 2018, January. The Neglected Social Dimensions to a Vehicle-to-Grid (V2G) Transition: A Critical and Systematic Review. Environmental Research Letters 13 (1): 013001, 1–18.

13 Sovacool, B. K., J. Kester, L. Noel, and G. Zarazua de Rubens. Contested Visions and Sociotechnical Expectations of Electric Mobility and Vehicle-to-Grid Innovation in Five Nordic Countries. Environmental Innovation and Societal Transitions, in press 2019.

14 Hjerpe, M., and B.-O. Linner. 2009. Utopian and Dystopian Thought in Climate Change Science and Policy. Futures 41 (4): 234–245.

15 Van Lente, H. 2000. Forceful Futures: From Promise to Requirement. In Brown, N., Rappert, B., and Webster, A. (Eds.): Contested Futures: A Sociology of Prospective Techno-Science (Aldershot; Burlington, VT: Ashgate), pp. 43–63.

16 Van Lente, H. 2012. Navigating Foresight in a Sea of Expectations: Lessons from the Sociology of Expectations. Technology Analysis & Strategic Management 24 (8): 769–782.

17 Ibid.

18 Berkout, F. 2006, July–September. Normative Expectations in Systems Innovation. Technology Analysis & Strategic Management 18 (3/4): 299–311.

19 Michael, M. 2000. Futures of the Present: From Performativity to Prehension. In Brown, N., Rappert, B., and Webster, A. (Eds.): Contested Futures: A Sociology of Prospective Techno-Science (Aldershot; Burlington, VT: Ashgate), pp. 21–39.

20 California Independent System Operator. 2014. California Vehicle-Grid Integration (VGI) Roadmap: Enabling Vehicle-Based Grid Services (Folsom, CA: California Independent System Operator).

21 Ibid.

22 Sovacool, Axsen, and Kempton (2017: 377–406).

23 Kempton, W., and S. Letendre. 1997. Electric Vehicles as a New Source of Power for Electric Utilities. Transportation Research 2 (3): 157–175.

24 Letendre, S. E., and W. Kempton. 2002, February 15. The V2G Concept: A New Model for Power? Public Utilities Fortnightly, pp. 16–26.

25 Kempton, W., and J. Tomic. 2005. Vehicle-to-Grid Power Fundamentals: Calculating Capacity and Net Revenue. Journal of Power Sources 144: 268–279. doi: 10.1016/j. jpowsour.2004.12.025.

26 Kempton, W., and J. Tomic. 2005. Vehicle-to-Grid Power Implementation: From Stabilizing the Grid to Supporting Large-Scale Renewable Energy. Journal of Power Sources 144: 280–294. doi: 10.1016/j.jpowsour.2004.12.025.

27 Hidrue, M. K., and G. R. Parsons. 2015. Is There a Near-Term Market for Vehicle-to-Grid Electric Vehicles? Applied Energy 151: 67–76.

28 Tomic, J., and W. Kempton. 2007. Using Fleets of Electric-Drive Vehicles for Grid Support. Journal of Power Sources 168: 459–468.

29 Guille, C., and G. Gross. 2009. A Conceptual Framework for the Vehicle-to-Grid (V2G) Implementation. Energy Policy 37: 4379–4390.

30 Yamagata, Y. et al. 2014. Energy Resilient Smart Community: Sharing Green Electricity Using V2C Technology. Energy Procedia 61: 84–87.

31 Hosseini, S. S. et al. 2014. A Survey on Mobile Energy Storage Systems (MESS): Applications, Challenges and Solutions. Renewable and Sustainable Energy Reviews 40: 161–170.

32 Hein, R. et al. 2012. Valuation of Electric Vehicle Batteries in Vehicle-to-Grid and Battery-to-Grid Systems. Technological Forecasting & Social Change 79: 1654–1671.

33 Debnath, U. K. et al. 2014. Quantifying Economic Benefits of Second Life Batteries of Gridable Vehicles in the Smart Grid. Electrical Power and Energy Systems 63: 577–587.

34 Arslan, O., and O. E. Karasan. 2013. Cost and Emission Impacts of Virtual Power Plant Formation in Plug-In Hybrid Electric Vehicle Penetrated Networks. Energy 60: 116–124.

35 Saber, A. Y., and G. K. Venayagamoorthy. 2010. Intelligent Unit Commitment with Vehicle-To-Grid—A Cost-Emission Optimization. Journal of Power Sources 195: 898–911.

36 Habib, S. et al. 2015. Impact Analysis of Vehicle-to-Grid Technology and Charging Strategies of Electric Vehicles on Distribution Networks—A Review. Journal of Power Sources 277: 205–214.

37 Lund, H., and W. Kempton. 2008. Integration of Renewable Energy into the Transport and Electricity Sectors through V2G. Energy Policy 36: 3578–3587.

38 Kester, J. et al. 2018. Promoting Vehicle to Grid (V2G) in the Nordic Region: Expert Advice on Policy Mechanisms for Accelerated Diffusion. Energy Policy 116: 422–432.

39 Sovacool, B. K. et al. 2018. The Demographics of Decarbonizing Transport: The Influence of Gender, Education, Occupation, Age, and Household Size on Electric Mobility Preferences in the Nordic Region. Global Environmental Change 52: 86–100.

40 IEA, Nordic Energy, 2018. Nordic EV Outlook 2018 105.

41 Berkout (2006).

42 Van Lente, H. 1993. Promising Technologies: The Dynamics of Expectations in Technological Developments (Enschede, NL: Twente University).

43 Bakker, S., H. V. Lente, and M. Meeus. 2011. Arenas of Expectations for Hydrogen Technologies. Technological Forecasting & Social Change 78: 152–162.

44 Van Lente, H. Seminar presentation. 2016, January 15. How Expectations Matter: The Sociology of Expectations and the Development of Needs (Copenhagen).

45 Deuten, J. J., and A. Rip. 2000. The Narrative Shaping of a Product Creation Process. In Brown, N., Rappert, B., and Webster, A. (Eds.): Contested Futures: A Sociology of Prospective Techno-Science (Aldershot; Burlington, VT: Ashgate), pp. 65–86.

46 Van Lente (2016).

47 Deuten and Rip (2000).

48 Berkout (2006).

49 van Rijnsoever, F. J., L. Welle, and S. Bakker. 2014. Credibility and Legitimacy in Policy-Driven Innovation Networks: Resource Dependencies and Expectations in Dutch Electric Vehicle Subsidies. The Journal of Technology Transfer 39: 635–661.

50 Borup, M. et al. 2006. The Sociology of Expectations in Science and Technology. Technology Analysis & Strategic Management 18 (3–4): 285–298.

51 Eames, M. et al. 2006. Negotiating Contested Visions and Place-Specific Expectations of the Hydrogen Economy. Technology Analysis & Strategic Management 18 (3–4): 361–374.

52 Van Lente (2000).

53 Ibid.

54 Bakker et al. (2011).

55 Borup et al. (2006).

56 Brown, N., A. Rip, and V. L. Harro. Expectations in & about Science and Technology. A Background Paper for the "Expectations" Workshop of 13–14 June 2003.

57 Van Lente, H., and A. Rip. 1998. Expectations in Technological Developments: An Example of Prospective Structures to Be Filled in by Agency. In Disco C., and van der Meulen. B. (Eds.): Getting New Technologies Together. Studies in Making Socio-technical Order (Berlin: Walter de Gruyter), pp. 203–231.

58 Ibid.

59 Bakker et al. (2011).

60 Ibid.

61 Berkout (2006).

62 Rappa, M. A., and K. Debackere. 1992. Technological Communities and the Diffusion of Knowledge. R & D Manage 22 (3): 209–220.

63 Bakker et al. (2011).

64 Ibid.

65 Wade, J. 1995. Dynamics of Organizational Communities and Technological Bandwagons —An Empirical Investigation of Community Evolution in the Microprocessor Market. Strategic Management Journal 16: 111–133.

66 D'Agostino, S. 1993, February. The Electric Car: A Historical Survey on the Motives Driving Its Existence (Institute for Electrical and Electronics Engineers Potentials), pp. 28–32.

67 Prud'Homme, R. 2010. Electric Vehicles: A Tentative Economic and Environmental Evaluation (International Transport Forum at the OECD: Discussion Paper 22).

68 Volti, R. 1990. Why Internal Combustion? Invention and Technology 6 (2): 42–47.

69 Schiffer, M. B. 1994. Taking Charge: The Electric Automobile in America (Washington, DC: Smithsonian Institution).

70 Hard, M., and A. Knie. 2001. The Cultural Dimensions of Technology Management: Lessons from the History of the Automobile. Technology Analysis & Strategic Management 13 (1): 91–102.

71 Smil, V. 2010. Energy Myths and Realities: Bringing Science to the Energy Policy Debate (Washington, DC: AEI Press).

72 Ibid.

73 Mom, G. P. A. 2004. The Electric Vehicle: Technology and Expectations in the Automobile Age (Baltimore, MD: Johns Hopkins University Press), p. 205.

74 Ibid.

75 D'Agostino (1993).

76 Cowan, R., and S. Hulten. 1996. Escaping Lock-In: The Case of the Electric Vehicle. Technological Forecasting and Social Change 53: 61–79.

77 Deluchi, M. et al. 1989, May. Electric Vehicles: Performance, Life-Cycle Costs, Emissions, and Recharging Requirements. Transportation Research Part A: General 23 (3): 255–278.

78 Rajashekara, K. 1993. History of Electric Vehicles in General Motors. Conference Record of the 1993 IEEE Industry Applications Conference Twenty-Eighth IAS Annual Meeting, Toronto, ON, pp. 447–454.

79 Chan, C.C. 1993. Present Status and Future Trends of Electric Vehicles. 2nd International Conference on Advances in Power System Control, Operation and Management, APSCOM-93, pp. 456–469.

80 Dijk, M., and M. Yarime. 2010. The Emergence of Hybrid-Electric Cars: Innovation Path Creation through Coevolution of Supply and Demand. Technological Forecasting & Social Change 77: 1371–1390.

81 Cowan and Hulten (1996).

82 Dijk and Yarime (2010).

83 Geels, F. W., G. Dudley, and R. Kemp. 2012. Findings, Conclusions and Assessments of Sustainability Transitions in Automobility. In Geels, F., Kemp, R., Dudley, G., and Lyons, G. (Eds.): Automobility in Transition? A Socio-Technical Analysis of Sustainable Transport (London: Routledge), pp. 335–373.

84 Melton, N., J. Axsen, and D. Sperling. 2016. Moving beyond Alternative Fuel Hype to Decarbonize Transportation. Nature Energy 1: 16013.

85 Dijk and Yarime (2010).

86 Orsato, R. J. et al. 2012. The Electrification of Automobility. The Bumpy Ride of Electric Vehicles towards Regime Transition. In Geels, F., Kemp, R., Dudley, G., and Lyons, G. (Eds.): Automobility in Transition? A Socio-Technical Analysis of Sustainable Transport (London: Routledge), pp. 205–228.

87 Melton et al. (2016).

88 Broto, V. C. 2012. Environmental Conflicts, Research Projects and the Generation of Collective Expectations: A Case Study of a Land Regeneration Project in Tuzla, Bosnia and Herzegovina. Public Understanding of Science 21 (4): 432–446.

89 Van Lente (2016).

90 Ibid.

91 Borup et al. (2006).

92 Michael (2000: 35).

93 Brown et al. (2003).

94 Pieri, E. 2009. Sociology of Expectation and the E-Social Science Agenda. Information, Communication, and Society 12 (7): 1103–1118.

95 Sovacool, B. K., J. Kester, L. Noel, and G. Zarazua de Rubens. Contested Visions and Sociotechnical Expectations of Electric Mobility and Vehicle-to-Grid Innovation in Five Nordic Countries. Environmental Innovation and Societal Transitions, in press 2019.

96 Walsh, P. R. 2012. Innovation Nirvana or Innovation Wasteland? Identifying Commercialization Strategies for Small and Medium Renewable Energy Enterprises. Technovation 32: 32–42.

97 Wentland, A. 2016. Imagining and Enacting the Future of the German Energy Transition: Electric Vehicles as Grid Infrastructure. Innovation: The European Journal of Social Science Research 29 (3): 285–302.

98 Müller, M. O. et al. 2011, October. Energy Autarky: A Conceptual Framework for Sustainable Regional Development. Energy Policy 39 (10): 5800–5810.

99 Michael (2000).

8

CONCLUSION

Dimensions, dichotomies and frameworks for energy futures

This book has examined the utopian visions, fantasies, frames, discourses, imaginaries, and expectations associated with six state-of-the-art energy systems—nuclear power, hydrogen fuel cells, shale gas, clean coal, smart meters, and vehicle-to-grid electric vehicles—playing a pivotal part in current social and policy debates about low-carbon policy, household practices, and decarbonisation. Theoretically, the book has evaluated these case studies with emerging concepts from various disciplines: utopianism (history of technology), symbolic convergence (communication studies), technological frames (social construction of technology), discursive coalitions (discourse analysis and linguistics), sociotechnical imaginaries (science and technology studies), and the sociology of expectations (innovation studies, futures studies). Spatially and contextually, we see visions with these energy systems at play globally in the two epistemic communities of nuclear physicists and hydrogen engineers as well as various public groups nationally in South Africa and the United Kingdom, and regionally in Eastern Europe and the Nordic countries.

Despite such a diversity of technologies, cases, and communities, there is a commonality to low-carbon fantasies and visions that is more systematically examined in this concluding chapter. Here, the empirical findings across each of the chapters is analyzed and synthesized, and the theoretical approaches are examined and compared. Empirically, visions tend to differ by problem addressed and function, storylines and cues, valence and contradiction, experiential subject, temporality, and degree of change. Conceptually, approaches can be juxtaposed in terms of their emphasis on structure, agency, or meaning, as well as their underlying assumptions. These conceptual and empirical differences can be mapped onto distinct dichotomies. The chapter ends by teasing out the book's broader conclusions and policy implications.

Dimensions and dichotomies of energy visions

The distinct visions or fantasies examined across the book—the five utopian visions of small modular nuclear reactors, five hydrogen fantasies, six frames connected to shale gas, six discursive coalitions and two themes with clean coal, nine imaginaries with smart meters, and the eight visions and expectations with electric mobility—add up to 41 in total. These visions seem to differ based on their problems and functions, storylines and characters, valence, experimental subject, temporality and degree of change. Each of these dimensions are explored in turn.

Problems and functions

At their core, each particular vision is pegged at some level to addressing some sort of problem—it therefore has a functionality or utility. This confirms earlier research which has suggested that visions and fantasies are often *functional* by fulfilling some perceived social need, and by enabling proponents to capture resources.[1] Expectations and promises about a new technology are thus "constitutive" and "performative" in "attracting the interest of necessary allies (various actors in innovation networks, investors, regulatory actors, users, etc.) and in defining roles and in building mutually binding obligations and agendas."[2] Such expectations serve to broker relationships between relevant social groups and create a dynamic of "promise and requirement" where actors make promissory commitments to the technology, forging a shared agenda that requires action. In this way, the functionality of the vision results in a "mandate" to developers and advocates: "the freedom to explore and develop combined with a societal obligation to deliver in the end."[3]

For instance, Eames et al. show how visions of a hydrogen economy draw touch upon six overarching problems:[4]

- Ending dependence on insecure supplies of energy;
- Decentralizing energy via community ownership of energy systems and smaller and more distributed sources of supply;
- Fundamentally reforming social values towards sustainability;
- Allowing humanity to retain its current lifestyles;
- Harnessing technical progress, knowledge, and innovation;
- Creating employment and staying in the international race for economic competitiveness.

When framing hydrogen as a solution to these problems, Eames et al. note that such visions intentionally keep the boundaries and possibilities limitless, enhancing rhetorical appeal.

I see similar dimensions, functions, and problems across all of the 41 visions depicted earlier in this book. As Table 8.1 indicates, some visions center on technological or scientific problems or ideographs such as progress, innovation,

TABLE 8.1 Frames, visions, narratives, and imaginaries with low-carbon technologies

Dimension	Frequency	Examples	Ideographs
Technological and scientific	N=7	Inevitability (clean coal and hydrogen), space exploration (nuclear), inclusive innovation (smart meters), costly disaster (smart meters), rapid electric society (electric mobility), innovation nirvana (electric mobility)	Progress, innovation, exploration, development
Socioeconomic	N=9	Economic opportunity (shale gas), economic sellout (shale gas), economic development (clean coal), maldevelopment (clean coal), poverty and racism (clean coal), progressive growth (hydrogen), astronomical bills (smart meters), families in turmoil (smart meters), broken and bankrupt businesses (electric vehicles)	Poverty, employment, economic growth
Environmental	N=9	Environmental boon (shale gas), environmental bane (shale gas), environmental sustainability (clean coal), environmental degradation (clean coal), environmental nirvana (nuclear), water security (nuclear), energy conscious world (smart meters), low-carbon grid (smart meters), ubiquitous and clean automobility (electric mobility)	Sustainability, futurity, stewardship, efficiency
Security	N=7	National security (shale gas), energy security (clean coal), risk free energy (nuclear), big brother (smart meters), hacked and vulnerable grid (smart meters, electric mobility), frozen families (electric mobility)	National security, safety, privacy, terrorism
Political	N=8	Authoritarianism (shale gas), indigenous self-energization (nuclear), energy independence (hydrogen), patriotism (hydrogen), energy democratization (hydrogen), empowered consumers (smart meters), energy autarky (electric mobility), captive consumers (electric mobility)	Liberty, democracy, empowerment, decentralization, independence
	N=41		20

Source: Author

scientific exploration, or technical development. Some center on socio-economic dimensions such as poverty, jobs, and growth. Some focus on environmental concerns such as sustainability, futurity, stewardship, or energy efficiency. Some involve security in various forms—national security, human security, safety, individual privacy, and terrorism. Some relate to politics and governance—liberty, democracy, empowerment, decentralization, and independence come to light.

Table 8.1 also brings three other things to mind: there is a fairly even distribution across technological and scientific, socioeconomic, environmental, security, and political dimensions, with the lowest frequency seven visions per dimension, and the highest nine visions per dimension. Second, there is an almost even distribution across technologies—all energy systems have visions that fall into at least five of the six dimensions. Lastly, the ideographs add up to 20 in total, suggesting a commonality to the deeper narratives or master frames behind each of the visions. The ideograph of progress in particular is able to give each group what they ostensibly want: a world of pollution free, and therefore limitless, energy use.

Furthermore, the dynamism between specific problems and solutions implies a "rhetorical selectivity"[5] that can entail positioning a future transition as unique so that comparisons with previous technological transitions can be overlooked (since the impact of this particular technological utopia will be so great that prior situations have no relevance), and ignoring the paradox of relying on technology to solve problems brought about by earlier forms of technology. In the extreme, such fantasies can be about deeming everything in the present as disposable, what David F. Noble calls an escapist sort of "technological pursuit of salvation."[6] Gabrielle Hecht's research is relevant here as well, for she has shown how maintaining nuclear technopolitical systems—what she calls nuclearity—requires "regimes of perceptibility" that generate invisibilities and erase conceptions from the past.[7] Human motivation pushes advocates to seek perfect visions, and since advocates frame problems in ways that advance their solutions as perfect choices, audiences remain limited in the questions they can ask because their acceptance of the problem orients them to accept a proposed solution. Thus, new technologies become evaluated primarily as solutions to existing problems (as defined by the technology's advocates).[8]

Storylines, symbolic cues and characters

Although less explicit and detailed, the visions throughout the book also differ in terms of their stories and narratives, plotlines, cues, and characters. Sometimes, these agents play active and conscious roles, at other times passive or even subconscious/unknowing roles. Some may be essential to securing a particular type of future, acting as a "star" or "lead actor," others may be important but not critical (acting as "extras" or "costars").[9]

Even though I didn't code for them, numerous typifications of actors and agents come up in throughout my chapters. There is:

- The product or artifact: the technology or sociotechnical system with limitless potential;
- The visionary prophet: the brilliant inventor, entrepreneur, or hero who seeks to selflessly better society with their innovation and effort;
- The happy user or consumer: the likely or intended adopter of the particular technology or service;
- The ally or intermediary: the critical stakeholder or champion whose support is needed for the innovation or technology to succeed;
- The prime adversary: the single person, stakeholder group, institution, or policy environment that must be overcome or eliminated;
- The unsatisfied critic: the annoying (and unreflective) person or organization who incessantly whines and can therefore be ignored because they are always unhappy;
- The evil villain: the malicious and deceptive provocateur set to destroy the world who must be stopped at all costs;
- The inhuman opponent: faceless threats such as climate change, poverty, or human insecurity that must be thwarted.

More specifically, within the chapters some storylines or visions—low-carbon grids, environmental nirvana, environmental boon/bane, water security—center on faceless evils such as climate change, degraded habitats, or disruptions to ecosystem services. Other times narratives such as energy independence, innovation nirvana, captive consumers or energy autarchy revolve around institutions, politicians, energy conglomerates, and other firms. Other visions involve more specific actors or individuals such as the hackers and terrorists behind a hacked and vulnerable grid.

Almost every vision can be tied in some way to a villain as well: the villains in the themes of inevitability (for clean coal and hydrogen) are those, often through ignorance and self-centeredness, which doubt the legitimacy of a foregone and beneficial energy transition. The villains in the themes of independence or autarky (smart meters, hydrogen, electric mobility) are oil suppliers, energy companies and cartels such as OPEC that desire to raise energy prices. The villains in the themes of patriotism or economic competitiveness or boon (shale gas, hydrogen, clean coal) are those that seek to waste energy and select inefficient technologies. Environmentalists across almost all of the visions see big industrial emitters of greenhouse gases as culprits threatening the vitality of the climate, whereas industrial stakeholders may see environmentalists as villains seeking to constrain growth and place limits on industrialization.

The resonance of these stories, characters, and villains reminds us that the publics and audiences subscribing to a particular vision will develop or reuse code words, phrases, slogans, and themes. These can trigger previously shared fantasies, may refer to a geographical or imaginary place or the name of a persona, and they may arouse a range of other emotions. It also reminds us that humans are storytellers, and when they share a dramatization of an event, they make sense out of its complexity by creating a script or narrative to account for what happened. This act of telling a narrative enables groups of people to come to a cognitive convergence about that part of their common experience.

Valence and contradiction

A third way visions differ—already alluded to when discussing problems, functions, and ideographs—is their valence. Some visions are utopian, and tied to positive emotions such as hope, excitement, happiness, and even love. Others are dystopian, and tied to negative emotions such as fear, despair, boredom, and even hatred. As Michael notes:

> Representations of the future are from the outset engaged in a sort of pre-emptive argumentation over whether the projected state of affairs leads to good or bad ... Representations of a good future can be charged with 'talking up' the future—that is, the enunciation of a particularly positive future can generate 'optimism' (or bullishness in markets). Similarly, a slight hint of negativity regarding the future can precipitate panic.[10]

Visons can thus frame low-carbon innovations as harbingers of a utopian social order, a nirvana for innovation and technical development, a platform for automobility ubiquity, or a pathway towards environmental sustainability, energy decentralization, community control, and autarky. These starkly contrast with negative visions of families literally freezing to death, small businesses declaring bankruptcy, terrorists and hackers launching new sophisticated attacks on grids, and consumers held hostage to the whims of unsentimental corporate firms.

Table 8.2 categorizes the visions throughout the book according to the positive and negative valence. Admittedly, there are no negative visions for nuclear power and hydrogen because I was analyzing only positive ones. But the negative ones for shale gas, clean coal, smart meters, and electric vehicles do bely that the low-carbon transport future itself is simultaneously pregnant with opportunity and full of promise but also dark, despairing, and despondent. It also implies a dynamic dialectic in that the positive visions are often defined only by relation to the negative ones (they avoid them), and vice versa.

As is obvious, the diametrically opposed valence of some visions does mean that multiple visions contradict each other. For small modular nuclear reactors, given the chapter's focus was on utopian visions, there is less explicit contradiction. However, there is still tension between the vision of risk-free energy (reduction of waste requires transmutation with longer half-lives) and environmental nirvana. Similarly, efforts to make reactors safer would make them less

TABLE 8.2 Valence of low-carbon visions

Technology	Visions	
	Positive and utopian	*Negative and dystopian*
Small modular reactors	Space exploration, environmental nirvana, water security, risk-free energy, indigenous self-energization	–
Hydrogen	Inevitable destiny, progressive growth, energy independence, patriotism, energy democratization	–
Shale gas	Economic opportunity, environmental boon, national security	Economic sellout, environmental bane, authoritarianism
Clean coal	Inevitability, economic development, environmental sustainability, energy security	Poverty and racism, maldevelopment, environmental degradation
Smart meters	Inclusive innovation, energy conscious world, low-carbon grid, empowered consumers	Costly disaster, astronomical bills, families in turmoil, big brother, hacked and vulnerable grid
Electric vehicles	Rapid electric society, innovation nirvana, ubiquitous and clean mobility, energy autarky, captive consumers	Broken and bankrupt businesses, hacked and vulnerable grid, frozen families

Source: Author

affordable, eroding the potential for indigenous self-energization. For hydrogen, the theme of energy democracy sees communities taking back control over production and use, whereas a theme of patriotism instead sees control shifted to companies and national economic competitiveness enhanced. For shale gas in Eastern Europe, environmental boon is the literal opposite of environmental bane, and the gains from economic opportunity are offset by economic sellouts. For clean coal in South Africa, economic development is pitted against maldevelopment, environmental sustainability against degradation, energy security against poverty and racism. For smart meters in the United Kingdom, inclusive innovation is literally the opposite of costly disaster, and empowered consumers is the opposite of big brother. For electric mobility in the Nordic region, energy autarky is the direct opposite of captive consumers, and a severe tension exists between innovation nirvana (which sees the fruits of innovation cascading to all) and hacked and vulnerable grids (which sees innovation furthering the destruction of society).

In some cases, such contradictions can be strategic. Berkout suggests that "a degree of flexibility over the interpretation of a vision can widen its relevance to greater numbers of actors."[11] Some contradiction can relate to the manufactured ambiguity or flexibility of most fantasies: they need to be broad enough to enroll actors but vague enough to withstand criticism. In other cases, tension and contradiction can be a way of furthering the plot—of enhancing the performative effect of a story's climax, resolution, or failure to reach a resolution.[12] In still other cases, such contradictions can reflect internal denial and cognitive dissonance among articulators of the vision or its audience. When looking at the early history of innovations such as electricity, x-rays, and the telegraph, Simon suggests that the introduction of any potentially transforming technology creates a tension between desirable changes in day-to-day life and the anxiety that follows any step into the unknown.[13] Marshall adds that some visions and fantasies act as psycho-social defense mechanisms, intended to assuage political discomfort or social anxiety.[14] Here the contradictory nature of a vision is not a weakness, or an unintended byproduct, but its intended strength or purpose.

Experiential subject

A fourth way visions can differ is based on their experiential subject—the person, group, or entity *experiencing* the intended (or unintended) consequences or results of a prospective future. The "subject" of a vision can be an individual, a group, an institution or other heterogeneous collective, or something far less tangible such as "the public," "society," "a future state of affairs," an "ecosystem" or "future generations."[15] The sort of subject being imagined or deployed invites audiences to identify with it, and it also creates a responsibility to them—to save them from suffering, to promote the greater good, to "do something to help them."[16] Creating experiential subjects for visions also ties back into agency as well, for narratives often utilize the language of "us" or "we" acting in a purposive, controlling way to change the future. Lastly, the tension or dynamism between specific and vague resonates with the themes above of problems/functions as well as valence and contradiction: futures are connected also to differing degrees of determinacy and indeterminacy.[17]

Admittedly, categorizing the visions in this book by subject was more difficult than by problems, storylines, and valence. Still, at a general level, experiential subjects range from specific individuals or decentralized entities (people, households, communities) to more vague collectives and centralized entities (society as a whole, the broader public, the biosphere). Table 8.3 offers an illustrative (rather than complete) list of subjects throughout the earlier chapters.

The fluidity of these experiential subjects serves as a reminder that each of the visions associated with particular energy systems are aimed at particular audiences or publics. Donald C. Bryant once noted that rhetoric performs "the function of adjusting ideas to people and of people to ideas."[18] Put simply, effective advocates manipulate their messages to resonate with their audiences and also use messages to help move audiences to be receptive to ideas. Successful rhetoric must reflect what is important to those people; the variety of experiential subjects helps proponents adjust visions to audiences and vice versa.

TABLE 8.3 Experiential subjects of low-carbon visions

	Subjects	
	Specific entities	*Vague collectives*
Small modular reactors	Scientists, space enthusiasts, investors, concerned citizens (often living near existing or proposed nuclear facilities)	Energy suppliers and conglomerates, environmentalists, the nuclear industry
Hydrogen	Community energy groups, energy consumers/users	The global economy, future generations
Shale gas	Gas/energy consumers, entrepreneurs, small business holders, civil society activists	Future generations, Eastern European society, oil and gas companies, Russia
Clean coal	Employees within the energy sector, regulators, financiers, energy consumers, displaced persons	The climate, the environment, future generations
Smart meters	Families, energy consumers, hackers, terrorists	The national innovation system, the economy
Electric vehicles	Families, community groups, drivers, passengers	Robots and artificial intelligence, digital and ICT companies, energy and equipment suppliers

Source: Author

Temporality

As alluded to already in Chapter 7, visions can differ in their temporality, some are proximal whereas others are distant. More immediate or proximal visions would occur in a few weeks to a few years' time. Distant visions are far into the future—usually at least a decade away, possibly a century away.

Recognizing that many visions fall into more intermediate areas between proximal and distant, and that many others offer no discernible or falsifiable claim about temporality, Table 8.4 offers a cursory sketch of the most proximal and distant visions within the book. Although the visions with small modular reactors were less specific about temporality, one could intuit that risk-free energy and indigenous self-energization would occur well before the entire energy sector is transformed via environmental nirvana, water insecurity is eliminated, and space exploration is perfected. The same goes for hydrogen—reducing dependence on foreign sources of energy, patriotism, and energy democratization have an immediacy to them, whereas inevitable destiny, and progressive and limitless growth would ostensibly take decades to occur. For shale gas, the jobs, pollution, and displacement of more carbon intensive energy carriers would occur almost as soon as production begins; effects on national security, changes in political structure, or the selling off of assets would come much later. For clean coal, the jobs, environmental effects, displaced

TABLE 8.4 Proximal and distant temporality of low-carbon visions

Technology	Visions	
	Proximal (within the next decade)	*Distant (within at least a decade)*
Small modular reactors	Risk-free energy, indigenous self-energization	Space exploration, environmental nirvana, water security
Hydrogen	Energy independence, patriotism, energy democratization	Inevitable destiny, progressive growth
Shale gas	Economic opportunity, environmental bane, environmental boon	National security, authoritarianism, economic sellout
Clean coal	Economic development, environmental sustainability, poverty and racism, maldevelopment, environmental degradation	Inevitability, energy security
Smart meters	Energy conscious world, empowered consumers, astronomical bills, families in turmoil, hacked and vulnerable grid	Inclusive innovation, low-carbon grid, big brother, costly disaster
Electric vehicles	The rapid electric society, hacked grids, frozen families, and captive consumers	Ubiquitous automobility, broken businesses, innovation nirvana, energy autarky, and captive consumers

Source: Author

persons and impacts on poverty are occurring or would occur once Medupi becomes operational. Others such as the inevitability of a fossil-fueled South African regime or enhancements to national energy security would come later. With smart meters, some visions—energy conscious world, empowerment, hacking and privacy issues, astronomical bills, families in turmoil—relate to aspects that would occur as soon as households adopt and use them. Others—inclusive innovation, a fully low-carbon grid, mass surveillance under big brother, the conclusion that the endeavor will be a costly disaster—will take much longer and require the transformation or complete transition of entire electricity, energy or national systems. Finally, for electric vehicles, the rapid electric society, hacked grids, frozen families, and captive consumers would be fairly proximal; ubiquitous automobility and broken businesses more distant; and innovation nirvana, energy autarky, and captive consumers most distant.

An implicit element of visions with differing timescales and immediacy (or temporal remoteness) is that some benefits may be pitted against each other. As Brown et al. write, future expectations "may run in parallel with and contest each other, occupying different time-frames and carrying different interests."[19] Some visions concern benefits and risks, such as economic development or water contamination, that occur now, while others, such as eventually halting climate change or transformations of social or economic structure, will occur well into the future.

The benefits or impacts of visions also occur at different scales: things like labor, land, air, and human health impacts tend to be localized, whereas progressive growth, the elimination of poverty, solving climate change, and space exploration are national, international, or even extraterrestrial. Such complexity plays a forceful role in making visions contested—something touched upon later in this chapter—and it also implies that the particular array of costs and benefits will play out differently according to each vision but mediated by temporality and place. The future will likely hold even more diverse and divergent pathways than these.

Another issue of temporality—not explored here but useful for future research—relates to how fantasies or visions present now can shift over time. Visions may have another degree of temporality connected to history and previous individual or collective experiences (i.e., hype cycles over previous failed transitions stigmatizing particular technologies today, such as electric vehicles). Nye notes for instance how mass fantasies can shift, what was once sublime can become ordinary. The railroad initially was seen as a "liberating machine," summarized in the 1800s as "a miracle of science, art, and capital, a magic power by which the forest is thrown open, the lakes and rivers are bridged, and all Nature yields to man."[20] Perhaps there is a general trend to visions that begin with utopian or dystopian claims and expectations, move to normalization and eventual disillusionment, and then, after technologies are abandoned or evolve, nostalgia.

Degree of change

A final way visions meaningfully differ is in terms of whether the scope of sociotechnical change brought about by the vision is incremental, pragmatic, or conventional or instead radical, substantive, and utopian.[21] Incremental visions essentially see the future pretty much as similar to the present, taking fundamental or foundational conditions of now as the basis of their foresight. These visions may even seek to protect, extend, or entrench business as usual. This contrasts with visions that are more progressive, disruptive, substantive, or ends-oriented, in which society may differ in fundamental ways from how it exists now. Malcolm Eames and his collaborators, for instance, have shown how visions of a hydrogen economy had contrasting incremental versus radical undertones.[22] Some scientists and engineers promoted hydrogen on the grounds that it would radically reorient society to hold more ecologically sustainable values—essentially transforming the energy system—whereas others promoted it because it would enable us to continue business as usual, preserving and extending the energy system. As Edward Tenner writes, this tension reflects a deeper social dilemma: "the choice is between a material and artistic culture that reflects and even anticipates change and one that cushions the spiritual shocks of change."[23]

I see such dynamics at play within each of the chapters (see Table 8.5). Given their utopian flavor, the only nuclear vision that is incremental is that of risk-free energy, essentially using reactors as we do now but without the pesky risk of accidents or waste. All of the other nuclear visions are transformative:

TABLE 8.5 Degree of incremental or radical change of low-carbon visions

Technology	Visions	
	Incremental or protective	Radical or disruptive
Small modular reactors	Risk-free energy	Indigenous self-energization, space exploration, environmental nirvana, water security
Hydrogen	Energy democratization, energy independence	Patriotism, inevitable destiny, progressive growth
Shale gas	Economic opportunity, national security	Environmental bane, environmental boon, economic sellout, authoritarianism,
Clean coal	Economic development, maldevelopment, inevitability	Environmental sustainability, poverty and racism, environmental degradation, energy security
Smart meters	Energy conscious world, empowered consumers, astronomical bills, families in turmoil, big brother, costly disaster	Inclusive innovation, low-carbon grid, hacked and vulnerable grid
Electric vehicles	Rapid electric society, energy autarky, frozen families, broken businesses, and captive consumers	Ubiquitous automobility, innovation nirvana, and hacked and vulnerable grids

Source: Author

spreading energy across every community, colonizing outer space, or alleviating completely environmental pressures and limitations. The hydrogen visions of energy democracy or independence are incremental, in that they merely transfer control of energy to communities or minimize dependence on imported supplies—the energy systems would remain much as it does today. However, visions of patriotism, destiny, and progressive growth would see more sweeping changes to national or even global structure. For shale gas, visions such as economic opportunity or national security are essentially about protecting the status quo—jobs, taxation, growth, energy independence—whereas visions such as environmental boon or bane, economic sellout, and authoritarianism are about more radical changes to the climate, local economy, or political structure. For clean coal, visions of economic development and maldevelopment are about the dangers of continuing business as usual (seeing Medupi as a catalyst or impediment), whereas visions such as sustainability, poverty, and degradation are about more serious—and at times irreversible—shifts in pollution or inequality. Incremental smart meter visions emphasize the household dynamics and interactions of smart meters—whether positive changes (energy conscious world, empowered consumer) or negative ones (higher bills, added family stress, intrusions on privacy, collapse of the program), the world would remain much as we know it. This contrasts with the more radical nature of a truly inclusive and innovative economy, a completely decarbonized grid, and a grid prone to constant hacking and terrorism. A radical vision for electric mobility would be one that depicts completely new forms of mobility such as robots, or flying cars, whereas an incremental vision portrays mobility much as it is today with human drivers and normal vehicles. The rapid electric society, energy autarky, frozen families, broken businesses, and captive consumers are all incremental visions; ubiquitous automobility, innovation nirvana, and hacked and vulnerable grids (with new forms of terrorism) are radical.

Conceptual approaches, frameworks and methodological strategies

Apart from revealing different empirical attributes and dimensions, the conceptual devices utilized in each of the chapters has been intentionally disparate. This was intended to reflect the diversity in approaches across multiple disciplines, including history, science and technology studies, futures studies, foresight studies, communication studies, sociology, and psychology. Despite being perhaps opaque to the uninitiated, such theories can serve as important heuristic devices that enable researchers (and hopefully readers) to make sense of large amounts of data; such frameworks also offer roadmaps for how to carry out empirical research. Indeed proclamations for theoretical breadth can be read as a celebration of diversity and the idea that no single framework or approach has a monopoly on truth.[24] Theoretical promiscuity can also open up discussions over underlying assumptions and epistemologies that often remain hidden.

Theoretical emphasis

Theories across the entire social sciences generally fit into five categories of where they tend to "center" their unit of analysis: agency, structure, meaning, relations, and norms, as Figure 8.1 depicts.[25] Agency-centered theories prioritize the agency of people, individuals, households, or interpersonal decision-making processes as well as those focusing on organizations or stakeholders. The second set of theories focus on structure, such as the macro-social, technological, or political environment, as their preferred focus. They may also conceptualize structure broadly to include analyses of institutional structure (such as positions of firms in an industry) and relations between technological systems and the natural environment. A third set of theories focus on meaning or discourse, highlighting elements such as language, symbolism, narratives, performativity, rhetorical visions, and discursive coalitions. These first three categories represent the classic social theory triangle of agency, structure, and meaning— or the three 'I's of interests, institutions, and ideas. A fourth type however moves beyond this triangle to refer to theories that attempt to apply their focus across agency, structure, and meaning. These hybrid theories are relational or processual, and may emphasize social relations and interactions, but they also highlight the webs of

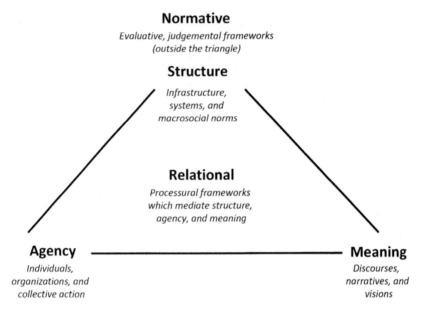

FIGURE 8.1   Agency, structure, meaning, relations and normativity in social theories
Source: Author, modified from Sovacool, B. K., and D. J. Hess 2017, October[28]

social structure and meaning in which actors are suspended and which they change through their action.[26,27] Finally, whereas the first four types of theories are inherently descriptive—describing people's agency, broader social or technical structure, language, or recursive relationships—a fifth type of theory attempts to answer whether a technology is a net positive or negative for society and individuals. To do so, they often rely on normative criteria set by ethics, moral studies, social justice or political ecology.

Under this framing, Table 8.6 shows how the six theories examined in the book fall loosely into three of these categories: agency, structure, and meaning. Technological frame and the Social Construction of Technology literature as a whole tend to emphasize the agency of relevant social groups, and how they attach meaning to different technological artifacts, leading to interpretive flexibility. Discursive coalitions focus more on the macro or structural constraints that discourse impose on humans—people become shaped to serve institutions and their discursive intentions. Technological utopianism, symbolic convergence, imaginaries, and expectations emphasize more the content of communication itself—and therefore focus primarily on how meaning is co-created through positive stories, rhetorical visions, imaginaries, and promises, expectations, and ideographs.

Methodological assumptions and epistemology

A second approach to classification begins with four ideal types of theory based on underlying methodological goals and assumptions, a typology modified from

TABLE 8.6 Description and emphasis of six conceptual approaches to energy visions and fantasies

Approach	*Disciplines*	*Key author(s)*	*Description*	*Emphasis*
Technological frame (SCOT)	Science and technology studies, history of technology	Trevor Pinch, Wiebe Bijker	The acceptance or rejection of new technical artifacts will depend on relevant social groups, their interpretive flexibility, and the achievement of things like closure and stabilization	Agency: relevant social groups determine meaning
Discursive Coalition	Linguistics, political science, discourse analysis, development studies, geography	Maarten Hajer, Arturo Escobar, Michel Foucault	A group of actors that, in the context of an identifiable set of practices, shares the usage of a particular discourse, a set of story lines over a particular period of time	Structure: discourses shape both agency and language
Technological utopianism	Science and technology studies, history, innovation studies	Howard Segal, David Nye, Frans Berkout, Carolyn Marvin	Large sociotechnical systems can inspire awe and feelings of utopianism and sublimity among users and supporters, and such emotions can be powerful forces shaping positive interpretations of new technologies	Meaning: utopian narratives dictate preferences
Symbolic Convergence Theory	Communication studies, psychology	Ernest G Bormann, John F. Cragan, and Donald C. Shields	Mass fantasies can form and chain out through symbolic cues, dramatis personae, and rhetorical visions	Meaning: communication determines fantasy themes
Sociotechnical Imaginaries	Science and technology studies	Sheila Jasanoff,	Collectively held, institutionally stabilized, and	Meaning: imaginaries motivate

(Continued)

TABLE 8.6 (Cont.)

Approach	Disciplines	Key author(s)	Description	Emphasis
		Sang-Hyun Kim	publicly performed visions of desirable futures, animated by shared understandings of forms of social life and social order attainable through, and supportive of, advances in science and technology	support and bind publics together
Sociology of Expectations	Innovation studies, sustainability transitions, future studies	Harro van Lente, Mads Borup, Nik Brown, S. Bakker	Science and technology are saturated with expectations, and are therefore prone to cycles of hype and disappointment	Meaning: expectations and visions drive innovation

Source: Author, drawn partly from Sovacool and Hess 2017: 703–750[29]

Gioia and Pitre[30] and further adapted by Sovacool and Hess.[31] This typology suggests that theories can fall into four main groups: functionalist-institutionalist, interpretivist, critical humanist, and conflict.

In this matrix of theories, an ideal *functionalist-institutionalist* theory would focus on systemic stability, equilibrium dynamics, organizations, and institutional settlements punctuated by periods of change and contestation. This theory type seeks to examine both regularities and changes that can be explained causally, and it seeks to develop theories to characterize structure and process. Structure may be understood as a relatively exogenous phenomenon that is external to, and independent of, agents but it is also produced and modified as actors contest normative and cognitive systems. From the list of theories above, the idea of stabilization and closure in SCOT approaches this ideal type.

Interpretivist theories are based on the view that people socially and symbolically construct and sustain their own realities, and the goal of theory is to generate thick descriptions, insights and explanations of the complex models of and for action that people share, reproduce, perform, contest and modify. Because this type of theory tends to focus on interpreting the meaningful social action, it tends to underplay the analysis of deeply transformative change. Symbolic Convergence Theory, the Sociology of Expectation and Technological Utopianism is consistent with the interpretivist type, but Sociotechnical Imaginaries and SCOT also have elements that approximate this type (e.g., imaginaries as cultural texts and the concepts of technological frames in SCOT).

A *critical humanist* theory, like the interpretivist type, also begins with social construction, discourse and meaning, but it draws more attention to historical contingency, cultural difference, and temporal and geographical contingency. This theory type also has a more critical or evaluative stance that enables researchers to develop insight about the underlying assumptions of the current social order by pointing to how they once were different, how they have changed over time, how they are different in other cultural or institutional settings, and how they are being imagined for the future. Doing so opens up the horizon of inquiry to appreciation of historical contingency, human agency and the potential for significant change and radical innovation. Discursive Coalitions and Sociotechnical Imaginaries, when implemented as historical and/or comparative projects, can approximate this type.

Conflict theory is related to the critical humanist type because of its focus on examining underlying assumptions in order to free the imagination to envision societal change; however, unlike critical humanist theories, the focus is more on patterns of structural inequality and institutionalized disparities. The goal of developing generalizable theory also suggests similarity with the functionalist-institutional type, but the conflict type draws attention to systemic sources of instability and conflict rooted in longstanding and durable structures of inequality. Because of the asymmetry of social conflict, theories of this type assumes that societal elites tend to control the means of production (and reproduction) of material, social, and cultural worlds. The goal of theory is to reveal, explain, and criticize existing mechanisms of domination. Discursive Coalitions approximate this type when they attempt to deconstruct or critique hegemonic narratives.

The concept of ideal types allows us to compare and contrast theories by identifying underlying assumptions and goals. Some theories may align with more than one type, whereas others may approximate more closely a single ideal type. Theories can, of course, bring together more than one type operating in a "transmission zone." For example, Discourse Theory can focus on radical change and deconstructive criticism associated with the critical humanist type but it can also involve cultural interpretation. Likewise, Technological Utopianism can blur the line between critical humanist and interpretivist approaches whenever it both describes and critiques. Figure 8.2 attempts to visualize these transmission zones and placement of theories, with the understanding that the placement in the figure does not mean that elements of other types are absent. The figure also emphasizes how only a few theories reside solely within a single ideal type.

Limitations and future research

Although this book has sought to be empirically rich, theoretically inclusive, topically and technologically diverse, and comparative in its exploration of case

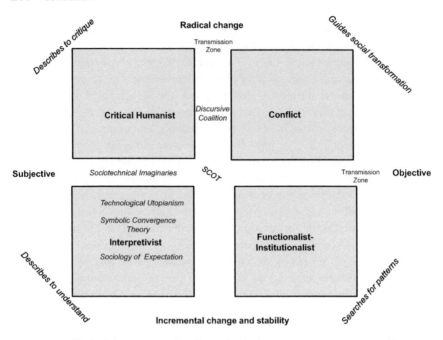

FIGURE 8.2 Underlying goals and epistemological assumptions to energy vision and fantasy theories

Source: Author, modified from Sovacool, B. K., and D. J. Hess 2017, October[32]

studies, it certainly has a number of important limitations that also point the way towards future research.

A most obvious gap in coverage concerns other low-carbon energy systems ranging from new state of the art designs for renewable electricity such as wind energy (onshore, offshore, and even cutting edge such as high altitude or vertical access), solar energy (photovoltaics, concentrated, thermal as well as cutting edge such as organic or extraterrestrial microwave), and algal biofuel (made in raceway ponds or photobioreactors). Other emerging technologies with both their own potential to reduce carbon and discursive dynamics are fusion, self-driving cars, negative emissions technologies and climate and geo-engineering systems.

A related gap concerns the geographic focus of the case studies examined. Analysis could and should be extended well beyond the handful of countries here, notably those in the Global South (especially least developed countries) or entire continents and cultures missed in this book, such as the Asia-Pacific, Middle East, South America, and the rest of Africa beyond South Africa.

The chapters here have explored a specific low-carbon technology, and its associated visions in isolation, but other work could look at competition between technologies as well as direct competition (or co-evolution) between their visions. Or, put another way, how multiple technologies and sociotechnical systems share the same vision or visions.

Also, given that it has sought most to reveal a heterogeneity of visions, the book has mostly treated actor groups or discursive coalitions as static (unchanging over time) and homogenous (hydrogen engineers, Nordic experts) when they are in fact diverse. Relatedly, given its focus on language and meaning rather than agency or structure, throughout the book cultural and historical comparisons are seldom made in any detail. How does national setting matter? What are the main aspects of national contexts steering imagination and narratives about energy futures? These important questions remain unanswered.

Likewise, given much of the material is prospective, looking at energy visions and forecasts, more information about the technologies themselves as they begin to emerge and become commercialized (and accrue better performance data, real world user experiences, cost profiles, etc.) would be useful to test and validate the visions here.

Another compelling line of inquiry concerns directionality. The book has hopefully done an exhaustive job identifying and classifying the 41 visions sprinkled throughout the chapters but it has not looked at causality or circulation. Where do these visions or fantasies begin—which actor groups or institutions invent or deploy them, or in which private, public, or professional spheres of communication do they occur—before being birthed or chained out and repro-duced more broadly? Do they germinate privately before proliferating into public discourse and impacting national policy (and programming and funding efforts)? Or do they begin in a national laboratory somewhere—or a novel, a diary, a myth or story that circulates—before securing funding and then chaining down to public discussion that affects individuals? Or (more likely) is the pattern of directionality unique for each innovation, bounded in technology, time, actors, and place? If so, how can we study (and derive generalizable and meaningful findings) from such specificity?

Conclusions and policy implications

Despite these nagging limitations yet imperative questions, the first and most wide-ranging conclusion from the book is that energy policy decisions made by analysts, politicians, users, scientists, and other stakeholders are not always purposively rational. I see 41 visions and 20 ideographs circulating across a mere six low-carbon energy sources (small modular reactors, hydrogen fuel cells, shale gas, clean coal, smart meters, and electric vehicles). The prevalence of these visions strongly suggests that current discussions and broader narratives about energy technology and policy seamlessly intertwined with compelling, and exciting and at times despondent and dystopian, fantasies. Visions about energy systems transcend economic self-interest, logic, and rationality and involve elements as diffuse as individual optimism, dramatic storylines, sym-bols, communal hope, business ambition, national pride, and fear of uncer-tainty. This offers a robust critique of disciplinary approaches to energy and low carbon transitions rooted in techno-centrism or determinism (focusing only on the technology) or economics (focusing only on prices and predicted behavior among actors). Public policy in particular must heighten awareness of non-logical (and non-economic) expectations in energy decision-making.

Because low-carbon fantasies fulfill these deeper needs, their provocative force can, in the right social and technical circumstances, play a meaningful role in how societies allocate resources relating to energy research and embrace (or reject) particular forms of energy conversion, use, and technology. It can lead to cycles of hype (inflated expectations) and disappointment (stigma) that do impact investment decisions, research trends, and socio-technical pathways.

Worryingly, visionary thinking can also distort a dispassionate discussion of risk, costs, and benefits. There is a danger to visionary thinking if it is true that the newest or most fantastic energy technology may not truly be the best one for society when all of its costs and benefits are coldly calculated. New technologies, by their very nature, tend to have more unknown and unforeseen risks than older, time-tested technologies. To promote and even oversell futuristic, exciting, revolutionary technologies has the potential to divert resources away from more cost effective solutions. As an important corollary to this point, new technologies —from wind turbines to electric appliances—often suffer from the same sorts of impediments, including lack of consumer information, suitable sales channels, barriers to entry, higher capital costs, and so on. To continually promote an entirely new set of technologies (instead of relying on the older ones) may fulfill our psychological needs for optimism and fantasy but may also mean these systems must overcome each of these impediments all over again, whereas more established systems and practices have already overcome them to gain social acceptance.

As Amory Lovins has so elegantly written, the energy problem is not how to expand supplies to meet the needs of society but rather how to accomplish the social goals of providing energy elegantly and with a minimum amount of energy and effort, and preserving the social fabric that sustains our communities and our greater culture.[33] If he is right, a sound energy strategy involves far less abundance, control over nature, national pride, and profligate use, and far more efficiency and sensible restraint. Energy fantasies and exaggerated rhetoric can become particularly hazardous if they blind us to the realities of new energy sources by promising a golden tomorrow only by ignoring the stark and growing problems of today.

That said, in each of the six cases examined, the visions associated with energy technologies are highly responsive and positive, addressing some type of challenge or problem facing society. The six low-carbon technologies examined are depicted as capable of achieving things as diverse as repairing broken down family values (or resulting in families freezing to death), eliminating poverty and hunger (or worsening them), reasserting individual or local autonomy (or eroding it), or colonizing outer space. These optimistic, visionary, and inspirational qualities can motivate people to become enrolled in the fantasy together, or become involved in actively stopping a dark and dangerous future. Imagined futures are emotive and sometimes transformative. Some visions even predict sweeping, radical changes, underscoring how fairly incremental changes to technology (electric motor, grid interconnectivity, fuel cells) can lead to visionary storylines.

Furthermore, the imagined futures are creative and collective. These 41 visions have multiplied to a surprising number of people, including those in the developing world, and also expanded to a broad base of advocates from universities, government bodies, research laboratories, the media and the public. Visions are not located in any one individual or group, nor are they confined to a particular type of technology, and instead manifest themselves across countries and cultures. This implies that the provocative force of visions and fantasy can, in the right circumstances, transcend any specific location or institution. Also, the popularity and recurrence of visions do suggest that the broader low-carbon future remains an open ended idea, as well as one subject to mass appeal, capable of sustaining the public imagination.

Moreover, despite these promising aspects of visions, the imagined futures are contested and contradictory. Within these chapters, there remains no consensus, no master vision or ideograph, about low-carbon innovation or energy supply and use, what it can or should do, whether it is positive or negative, or how quickly it will occur. This only underscores the mixed emotive forces at play behind the visions: people may simultaneously feel excited and anxious. Because of this contestation, the more specific future of each of the six low-carbon technologies examined remains uncertain.

Nonetheless, ideographs remain constant and recurrent across my visions. Visions at a narrow level about a cleaner type of natural gas or coal, a smaller reactor, a more widely dispersed fuel cell or smart meter or vehicle come to symbolize much more than that. They borrow from and connect with deeper ideographs of progress, convenience, environmental sustainability, physical shelter, liberty, autonomy, privacy, security, safety, love, employment, and economic growth (to name a few). In this way, expectations and promises satisfy a general social need, and as such will likely continue to exist and evolve even as sociotechnical systems accelerate, stagnate, or collapse. The modern world may be alienating for many of us, and an underlying dissatisfaction with modernity and some essential causes behind it—such as workplace dissatisfaction, family disorder, community fragmentation, environmental decay, political powerlessness, and spiritual emptiness—have not been alleviated. Fantasies relating to mastery and control, utopian social orders, cheap sources of energy, and symbols of national pride (among others) perhaps provide a degree of comfort and excitement in this complex and at times depressing world. They can at times blind us to substantial obstacles with achieving a low-carbon future but in other times inspire and motivate us to overcome them.

Notes

1 Geels, F. W., and W. A. Smit. 2000. Failed Technology Futures: Pitfalls and Lessons from a Historical Survey. Futures 32 (9–10): 867–885.

2 Borup, M. et al. 2006. The Sociology of Expectations in Science and Technology. Technology Analysis & Strategic Management 18 (3–4): 285–298.

3 Borup et al. (2006).

4 Eames, M. et al. 2006. Negotiating Contested Visions and Place-Specific Expectations of the Hydrogen Economy. Technology Analysis & Strategic Management 18: 361–374.

5 Peterson, T. R. 1997. The Power of Ambiguity. In Sharing the Earth: The Rhetoric of Sustainable Development (Chapel Hill, NC: University of South Carolina Press), pp. 34–53.

6 Noble, D. F. 1997. The Religion of Technology: The Divinity of Man and the Spirit of Invention (New York: Penguin Books).

7 Hecht, G. 2012. Being Nuclear: Africans and the Global Uranium Trade (Cambridge, MA: MIT Press).

8 Burke, K. 1966. Language as Symbolic Action; Essays on Life, Literature, and Method (Berkeley, CA: University of California Press).

9 Brown, N., B. Rappert, and A. Webster. 2000. Introducing Contested Futures: From Looking into the Future to Looking at the Future. In Brown, N., Rappert, B., and Webster, A. (Eds.): Contested Futures: A Sociology of Prospective Techno-Science (Aldershot; Burlington, VT: Ashgate), pp. 3–20.

10 Michael, M. 2000. Futures of the Present: From Performativity to Prehension. In Brown, N., Rappert, B., and Webster, A. (Eds.): Contested Futures: A Sociology of Prospective Techno-Science (Aldershot; Burlington, VT: Ashgate), pp. 21–39.

11 Berkout, F. 2006, July–September. Normative Expectations in Systems Innovation. Technology Analysis & Strategic Management 18 (3/4): 299–311.

12 Deuten, J. J., and A. Rip. 2000. The Narrative Shaping of a Product Creation Process. In Brown, N., Rappert, B., and Webster, A. (Eds.): Contested Futures: A Sociology of Prospective Techno-Science (Aldershot; Burlington, VT: Ashgate), pp. 65–86.

13 Simon, L. 2005. Dark Light: Electricity and Anxiety from the Telegraph to the X-Ray (New York: Mariner Books).

14 Marshall, J. P. 2016. Disordering Fantasies of Coal and Technology: Carbon Capture and Storage in Australia. Energy Policy 99: 288–298.

15 Michael (2000: 21–39).

16 Ibid.

17 Brown, Rappert, and Webster (2000: 3–20).

18 Bryant, D. C. 1953. Rhetoric: Its Function and Its Scope. Quarterly Journal of Speech 34: 401–424.

19 Brown, Rappert, and Webster (2000: 3–20).

20 Nye, D. E. 1994. American Technological Sublime (Cambridge, MA: MIT Press), p. 46.

21 Michael (2000).

22 Eames et al. (2006: 361–374).

23 Tenner, E. 2006. Winter. The Future Is a Foreign Country. The Wilson Quarterly 30 (1): 62–66.

24 Sovacool, B. K., and D. J. Hess. 2017, October. Ordering Theories: Typologies and Conceptual Frameworks for Sociotechnical Change. Social Studies of Science 47 (5): 703–750.

25 Ibid.

26 Geels, F. W. 2009. Foundational Ontologies and Multi-Paradigm Analysis, Applied to the Socio-Technical Transition from Mixed Farming to Intensive Pig Husbandry (1930–1980). Technology Analysis & Strategic Management 21 (7): 805–832.

27 Rutherford, J., and O. Coutard. 2014. Urban Energy Transitions: Places, Processes and Politics of Socio-Technical Change. Urban Studies 51 (7): 1353–1377.

28 Sovacool and Hess (2017: 703–750).
29 Ibid.
30 Gioia, D. A., and E. Pitre. 1990. Multiparadigm Perspectives on Theory Building. Academy of Management Review 15 (4): 584–602.
31 Sovacool and Hess (2017).
32 Ibid., 703–750.
33 Lovins, A. 1976. Scale, Centralization, and Electrification in Energy Systems. In Future Strategies for Energy Development: A Question of Scale (Knoxville, TN: U.S. Department of Energy and Oak Ridge National Laboratory), pp. 88–171.

INDEX